Vital Media

Vital Media

Making, Design, and Expression for Humans and Other Materials

Michael Nitsche

The MIT Press
Cambridge, Massachusetts
London, England

© 2022 Massachusetts Institute of Technology

This work is subject to a Creative Commons CC-BY-NC-ND license.
Subject to such license, all rights are reserved.

The MIT Press would like to thank the anonymous peer reviewers who provided comments on drafts of this book. The generous work of academic experts is essential for establishing the authority and quality of our publications. We acknowledge with gratitude the contributions of these otherwise uncredited readers.

This book was set in Stone Serif and Stone Sans by Westchester Publishing Services. Printed and bound in the United States of America.

Library of Congress Cataloging-in-Publication Data

Names: Nitsche, Michael, 1968– author.
Title: Vital media : making, design, and expression for humans and other materials / Michael Nitsche.
Description: Cambridge, Massachusetts : The MIT Press, [2022] | Includes bibliographical references and index.
Identifiers: LCCN 2021061884 (print) | LCCN 2021061885 (ebook) | ISBN 9780262544580 (paperback) | ISBN 9780262372060 (epub) | ISBN 9780262372077 (pdf)
Subjects: LCSH: Design—Human factors. | Sustainable design.
Classification: LCC NK1520 .N58 2022 (print) | LCC NK1520 (ebook) | DDC 745.4—dc23/eng/20220518
LC record available at https://lccn.loc.gov/2021061884
LC ebook record available at https://lccn.loc.gov/2021061885

10 9 8 7 6 5 4 3 2 1

Contents

Acknowledgments   vii

1   **Troubles and Necessities**   1
2   **Mapping Vital Media**   17
3   **Performance Makers**   53
4   **Recentering**   109
5   **Digital Folk**   167

References   205
Index   217

Acknowledgments

This is a book about interdependencies and balance. It tells, again and again, stories of shared becoming. Maybe it is not surprising, then, that writing the book was itself an exercise in balance, and it would have been impossible without the help, voices, and encouragement of many contributors.

These include my amazing colleagues at the School of Literature, Media, and Communication at Georgia Tech and particularly the faculty in the Digital Media program. I am glad that I found an academic home this colorful and supportive. I am particularly grateful for Philip Auslander and Jay Bolter, who provided feedback on early sections of the book. Any academic unit is only as good as its students, and I am most indebted to all the students who worked with me over the years. This includes the members of the Digital World and Image Group, which has been the home for my research for many years. Among those students were Hye Yeon Nam, Andrew Quitmeyer, Friedrich Kirschner, Jihan Sherman, and Claire Stricklin. Their curiosity and enthusiasm remain infectious. Some of their amazing work will showcase in the book's argument, but working with them shaped my research at large and it led me into surprising venues that I do not want to miss.

Over the years that it took to assemble this project, I conducted numerous interviews with crafters, designers, and researchers about their work and their practices. Only a fraction of these conversations is included here, but I found all of these conversations inspirational and humbling. The ones that are covered in the book include Ehren Tool on his pottery practices, Cat Mazza on her work on the *Nike Blanket Petition*, Betsy Greer on craftivism, John Burrison on his folklore research, Steve Turpin and Mike Craven on their work as practicing folk potters, and Amit Zoran on his practice of hybrid craft. In addition, Phil Horzempa provided more details on the

Lunar Orbiter 1 section, and James Wagner Au provided details for the *Second Life* section. I owe many thanks to all of them, for their time, patience, and openness.

Many thanks are due to the team at MIT Press: to Doug Sery, who originally brought the early manuscript in, to Noah Springer, who turned that manuscript into an actual book, to Lillian Dunaj who helped get the book ready for editing and production, and to Cheryl Hirsch for her help during the editing process. They were supportive throughout and gently guided me through the daunting task of finishing this project. A particularly big thank-you is due to the anonymous reviewers who provided feedback along the way. Their responses were extremely helpful, challenging, and inspirational. Any shortcomings and errors of the book are entirely my fault, but their rich feedback was most valuable.

None of this work would have come into being without my loving family. They had to deal with my ramblings about spacecrafts and pottery and digital media for much too long. Without the help of family afar, in Germany and Poland, and nearby, in Atlanta, none of this would have happened or made sense. This rings most true to my family at home: Karolina, Klara, Nicholas, and Hannah. To them in love: thank you.

1 Troubles and Necessities

This book is about media and relationships. It sees media as dynamic networks where cognitive and noncognitive participants cocreate, and it calls for necessary adjustments in media design to balance the needs of all partners involved. Our current approach to mediating and positioning ourselves in the world according to a humancentric way is not sustainable. As a correction, I propose the ideal of vital media and argue that we need to reconsider our own position toward material stuff and media practices. Both of these pose challenges. On the one hand, the properties of that "stuff" have changed. Global networks, ubiquitous computing, and other forms of digital media have added new qualities to objects that once seemed content with themselves. We now find that our stuff speaks back to us in novel ways. This may express a new quality, but material conversations did not start with the digital. I will argue that materials always spoke and instead of focusing on the digital opportunities, I will make the counterargument that the voices of materials should not be overpowered or replaced by the digital. They are not ours or any computer's to control in the first place and do not require seamless remediation. Instead, I call for the careful inclusion of digital qualities in relation to material ones.

This strengthens the voices of materials, but the role of humans in the shaping of this stuff-world remains substantial. Human activity has irrevocably shaped the face of the world—a fact we recognize painfully slowly. We critically reflect on the Anthropocene through media. But are our media up for this task? I will propose the idea of vital media as an ideal to support a new balance in media design that depends on shared agency. Rebalancing mediation means not only that humans thrive but that all partners involved in mediating gain identity and diversity.

I argue from the crossroad of two interconnected realizations: The first is the story of the material turn. Things—living and nonliving, digital and nondigital—are being recognized as agents, and the way they intra-act in our media designs needs attention. The second is the story of the Anthropocene. Humans have become an evolutionary factor that has led to unsustainable conditions. I argue that if media center on humans alone, then—by design—they cannot break out of this cycle. This book presents the ideal of vital media to break out of that cycle. The first chapter outlines the basic problem, method, limitations, and structure of the argument to come.

What Could Possibly Be Wrong?

We are falling apart. At best, we face a double challenge. The first challenge concerns encounters with materials: their making, manipulation, and sustainability. How can we live in a world without destroying it? The second concerns the realization of ourselves: how we might imagine and reason. How do we continue without disassembling the human project? This is not a playground for academic debate but a daily challenge for everybody that will not be resolved by an invention or a unifying theory. The best possible outcome is not a fixed conclusion but to recognize that this has become an ongoing and shifting struggle.

Why write a book that will not provide a final equation, a neat model, or promises of future human glory but that instead reduces the role of the human, raises the voices of materials, and calls for humility in an ongoing balancing act? Is this not a massive letdown? On the contrary. The voice of this book is a hopeful one, and even though it might not provide any resolve, it still aims to deliver an alternative. It is a book that is fundamentally convinced of the importance of the human contributions to the world. But it is also a book that sees them in dependency to all other contributors.

Humans need to adapt to a new way, a way that no longer focuses on them alone. We have to make an existential change in the way we understand ourselves. The urgency of this is made clear enough in every media format available. We are using media to make meaning, debate ideas, and develop solutions. But what if the media we build our discussions on are flawed means themselves? What if they are part of the problem and not designed to open the necessary novel pathways? What if their available engrained mechanisms are not sustainable but increase or even accelerate the problems? Envisioning and criticizing media is the domain of media studies, and the shortcomings of media have been masterfully dissected by many great scholars. Yet, for the most part, these studies have been dominated by a humancentric logic themselves. We try to make media better for us. Sean Cubitt (2017, 5) talks of the "original sin that severed humans from their environments: the privilege granted to communication, a necessary survival mechanism that, however, has come to risk the survival of the whole ecosystem." He goes on to argue for "finite media" that depend on ecological environmental balance, which will require severe changes to the human way of life: "It is we ourselves who must become other in order to produce an other world. The correlative is that we must cease to be human, and most of all cease to exist as exclusively human polity, which is the medium of communication par excellence. The road to that goal, however, must lead through the polis, the humanity of humans, and most of all through our communications in order to imagine a way out of stasis" (2017, 5–6). This book is part of the struggle for such a fundamental change. It still sees a role for the human contribution and argues that what is so exceptionally human has to become part of a shared solution. Secondly, this book does not directly target environmental change as the main goal. Instead, it is limited to the material encounter and how it relates to the production of expression. Media are the junctions where the relationships between different partners form, where materials diversify and individuation happens. They form the crossroads where these two core challenges meet. It is the task of a media designer to structure this meeting point, to design its forming activities, to provide new opportunities, and to reimagine already existing ones. It is to those media designers that this book speaks as it talks about interaction and media objects and as it readjusts the reference of whose action we need to take into account.

Media designers are not content providers. They do not have to comply to existing formats and fill them with stories. Instead, they are tasked with improving the mediating operators overall and shaping their ecology. Sometimes the phrase *experience design* captures part of this task. But that phrase also emphasizes the fatal focus on the human "experience" instead of the wider ecology. As long as we design only for the experiences of humans, we remain trapped in the "original sin." The goal of this book is to inspire media designers in their efforts to renegotiate our media developments, to adjust them toward a balance that allows a continued emergence of all partners involved, including humans.

One way to construct such an argument would be to look back and dissect past fallacies identifying the mistakes made. This is not the path chosen here. Although the challenge of the material-social balance dominates the book, it will not trace its historic, social, or technological origins. Instead, this destructive imbalance is taken as a given and a point of departure to work from. From here we look forward and build on practical examples, as well as scholarly theories, to work toward a collection of helpful references. How can we achieve a balance between the needs we have for self-expression and individuation with the material needs of the world we live in? What criteria might define a media ecology that is also sustainable? From here, we arrive at the smaller yet still daunting question at the heart of this book: How can we build media that give voice to both the human and nonhuman? These are the media needed to support our dealings with the larger problem at hand. They are the tools necessary to achieve the new balance.

To answer this question and to build these tools, media design needs to be readjusted from a humancentric take to a new balance that acknowledges the role of materials and nonhuman participants. This in itself will not provide solutions for environmental problems. It will not reduce carbon footprints or waste. But it might offer ways to rethink media design in a way that inherently aims for material inclusivity. That inclusivity brings with it a range of new dependencies and qualities that support a better mediation of coexistence. These new media allow us to tackle the challenges before us in novel ways.

Forms of new materialism and feminist technoscience are guiding the path forward. Thankfully, they have already found their way into media studies (e.g., Bollmer 2019; Kember and Zylinska 2012) and design (e.g., Rosner 2018), and we can build on this material turn. This means that it is not necessary to revisit the foundational arguments for new materialism itself. We take it for granted that this world is not about humankind alone and that humans are but one component of its emergence. The first condition of this book is an acknowledgment of the new role of materials, and the foundations for this have been beautifully argued before (e.g., Barad 2007; Bennett 2010). Now that the human being has been repositioned as one among countless active ingredients in an increasingly flat hierarchy of agents that actively form our world, how can they find their place in these "media ecologies" (Fuller 2005)?

The second condition under which this question is asked is the role of digital media. Emerging media ecologies are infused with countless digital media, from server farms that support social media to personal devices that continuously track minute medical data. A core element provided by this digital dimension is procedurality. This quality is based on the understanding of computers not as "a wire or a pathway but an *engine*" designed to "embody complex, contingent behaviors" (Murray 1997, 72). Janet Murray's "engine" highlights the concept of the technical object and the forces it produces. Procedurality has been around since the first construction of engines capable of embodying certain processes, from the Jacquard loom to puppet-like automata. But the particular technological conditions of the digital have propelled the procedural into new dimensions. Artificial intelligence, ubiquitous computing, and global networks are some prominent features of the kinds of developments that affect material objects inside out. Digital operational logics have been added to countless new objects—from coffee cups, to notebooks, to shoes, to our bodies. Not only do humans grown into cyborgs but so do all other beings. Digital media pervade and provide novel operations to objects. They shape material objects whether digital components are embedded or not. The impact of digital networks on the production, distribution, and use of commercial objects is foundational. We might eat an apple that required a whole network of digital systems to be produced, packed, shipped, and sold. The digital is a quality that shapes objects that actively embody it as well as it affects those that do not. Thoreau's cabin at Walden Pond is largely gone, but its site can be visited via Google's street view. Digital media, such as complex management programs of fruit produce companies or Google's street view, change the way we perceive any apple or lake; through that they change their movements, growth, or destruction. Lakes and apples have changed. The digital is an evolutionary expansion of object and media qualities. Digital media change our perceptions *and* interactions with the world around us, but like everything else, they are still part of that world. The digital, like all other forces we have to deal with, accumulates to the cacophony of voices and actions that a media designer needs to handle. Because the digital remains part of the lived world, it does not offer any escape from the trouble we might have with sustainability, but it is part of it.

The path to navigate these two conditions is motivated by a troubling observation: the persistence of a continued separation between media worlds and "real" worlds. Many years ago, I had the opportunity to work in Sony's development studio in Cambridge, where I experimented on narrative spatial designs on the original PlayStation console. It was right before the launch of the PlayStation 2, and the fabled development kits were hidden away on a different floor. Hype surrounding the promises of the new technology was in the air. New virtual worlds were just around the corner to tickle our senses. These were the promises of alternate realities and such catchphrases as PlayStation 2's "Live in your world, play in ours." Many designs aspired to high levels of immersion (as they still do), and new media boldly stepped forward to offer the requested alternative worlds. The wonders were out there, and they were blissfully disconnected from the archaic, rigid conditions of the past. Around the launch of the original PlayStation, John Perry Barlow (1996) tried to claim independence for the emerging communities on the internet from the forces of governments and commerce: "Ours is a world that is both everywhere and nowhere, but it is not where bodies live." Yet the world was still "ours"; it was somebody's even though it was the technological new age. Technologies such as the internet or game consoles might evoke wonder and offer novel experiences. They certainly invite discoveries of new worlds, and my own interests at that time aimed at an optimization of these virtual spaces. But when the excitement of the new fades, then it is necessary to realize that the living body and its dependencies on the emerging world around it always remain at the heart of what media are about. That is why it is troubling to still find the very same logic at work a generation later when the commercial for the VR system Oculus Rift asks us to "defy reality." Putting up a virtual media world against the reality we live in clashes with the foundations of the argument presented here. Looking for procedurality in digital systems alone is outdated.

That does not mean that this book turns away from the digital. As noted above, digital qualities are a given. We cannot avoid them. The impact these forces have on our daily life and communities have been debated at length (e.g., Borgmann 1999; Castells 1996), but they are here to stay. The question is not one of unplugging our computers but of how to integrate digital design into wider media ecologies. This answer requires a focus on the material world, on a multiagent, critter-infested, flattening universe. It also calls for a turn away from "reality defying" alternative promises. The goal is not to take digital media out of the equation but to realize their own qualities within the necessary rebalance. There is no reason to construct concepts that sidestep or undercut a shared world. If there is no alternative state to escape to, then we need to design media that step away from a human-centered dead-end street and integrate nonhuman conditions better. We cannot continue to trust media that inherently lack such a balance to provide the means to step out of a cycle of material exploitation.

What humans bring to the ecology of media is their cognitive ability and their need for self-realization. Humans need to express themselves, and this is a key driving force for media in all their variety. Expression manifests differently in different cultures. At times, the individual's expression is emphasized; at other times, those of a social group, family, or tribe. But they are still means of self-realization in context. Because media ultimately respond to the human need for such an expression, this book will remain dependent on the human condition. It considers media as part of and a challenge to the human endeavor. The practices of that endeavor have to connect to the material world we are part of and address the challenges as the world that contains us has changed and flattened. Media add to the emergence of the world and—given the fatal errors we have committed in this regard so far—their rebalancing is a very necessary part of a much larger correction.

The approach centers not on a particular technology. It does not repair social media, artificial intelligence, big data, or any other technical domain. It is not an approach that will describe how to build or support new virtual communities or how to deal with networked societies. Instead, it will propose a quality—*vital media*—as a guiding ideal. To outline this quality, it will look at how individuals encounter material agencies. It looks at the moment where the hand touches the material—or is touched by it. Digital media support new actions in this encounter, but framing them as solutions is misleading. Aiming for a digitally driven redemption is the hopeless logic of the alternative world all over again. Instead of replacing worlds or even dominating them, the underlying argument is that media can only be relational. It is only logical that a key word that will follow us throughout this book is *balance*.

Limitations

First, we will focus on those media and materials that relate to the individual human body as we explore what the body touches and shapes and by what it is shaped in return. This close hands-on level poses challenges that differ from the grander scale of global media technology. In many ways, it stands in contrast to media that seem to celebrate individualism in the image of media-savvy content producers and "makers." Many social media platforms promise more access, more power to produce, more ways to distribute. But the very same media threaten to detach us from the material encounters that underlie those images or objects. This can produce a lack of identity while seemingly celebrating it at the same time. Our encounters with computers happen through standardized machines that promise ever more riches of personal communication as long as we follow their preestablished pathways, but we do not understand those inner operations. It threatens to dry out individual expressive richness in an ocean of conventions and lack of control. This not only applies to the condition of current social media but also to personal fabrication, video games, and other domains. The result is a widening gap between the ideal of a self-reliant maker and a dependent user. This gap is partially driven by commercial interests. The better the underlying technology is hidden, the more wonderous it performs, the more it simplifies the process, the more we are willing to pay for it, it seems. The more the conventions and standards are engrained, the easier they can be patented and exploited. The extinction of messy and difficult material experiences has become a symbol of efficiency and progress. It was a sign of the true gentleman of the sixteenth century to not depend on physical labor and have his servants in charge of material encounters, such as cleaning, cooking, or gardening. Today, there is a danger of perfect media designs to "gentleman" us from connections to the world we live in. We need to renegotiate these conditions.

The challenge is twofold: how to support expressive means for the human and, at the same time, support the role of the material. To address this dual challenge, we look into two main domains that are interconnected but offer differing vantage points. The first is performance as a prime example for the construction of personal and shared human expression. The second is craft as the creative practices between the maker and the object. Both fields are problematic. Performance rejects classification from its very inception and craft faces fundamental changes in practice and theory. Both cases require definitions that help to position one in relation to the other as we scope out a new path for media design across them.

One downside to the approach championed here is that it borders on the idiosyncratic. Large sections of this book ask how "I" can continue to create "my" "self" in the new conditions and how the voice of a material might emerge from that toward "me." Focusing on the individual should not be misunderstood as an egoistic or even human-centered perspective. Neither does it proclaim the ideal of the individual over their community. Centering on the individual encounter is a limitation based on scope and focus. Media shape the world as well as the person. This can take us into abstract territory, which answers more questions about the philosophical nature of our encounters than how we might deal with them. This book is not providing a new media philosophy but looks for applications in design. The focus remains on the individual's encounters with the material to provide media designers with examples, qualities, questions, and ideas for their own practices.

The goal is not to utilize or optimize media for any specific purpose. This is not a treatise on tactical media. It precludes such a tactical media perspective in some ways. Sarah Kember and Joanna Zylinska (2012, 177–178) critique a tactical media approach as they observe that "many tactical media 'hacktivists' seem to lose sight of the conditions of material and symbolic production that are involved in setting up such oppositions." Many artists and activists manage to make materials and means of production an integral part of their work but the challenge remains: before subscribing to a particular tactical approach, one first faces the foundational nature of material renegotiation, the "setting up." The following argument first explores this heart of media production. It roots them not in a cause but in the idea of vital media.

Vital media are an ideal that grows out of an evolutionary take on media design. This means that it itself cannot claim to be free from deeply flawed manifestations of the emerging media. Producing vital media might still lead to seriously flawed expressions. They might still tell stories of destruction, exploitation, and manipulation. These media, as all media, are still capable of producing toxic debates, propaganda, and advertisements. Yet vital media also provide arguments to critically reflect on their production and help to address destructive material conflicts through their design and operational processes. Including material operations in our thinking about media introduces them as checks and balances. The hope is that they provide us with an inherent survival mechanism through their foundations in materiality. Thus, vital media designers listen to more than our human voices. They do not build a hierarchy but allow the emergence of diverse perspectives through new media practices based on the needs of all partners involved. In that regard, the approach taken here might be described as participatory design with stuff that does not look for a "way out" of digital culture but for a "way in" via a material turn.

Methods and Aspirations

This book builds on the "material turn" and relates to what has been termed *nonrepresentational research* (Vannini 2015). It stands on the shoulders of foundational contributions by Karen Barad, Donna Haraway, and Bruno Latour, as well as Jane Bennett and Tim Ingold. The work of these scholars differs, conflicts, but often speaks to each other as they all emphasize the increased role of nonhuman actors in relation to the human ones. By and large, the argument here will not recount them. The material turn is taken as a given, and the question is how to move on to design media in that condition. In many ways, this book does not retrace the arguments of those scholars but builds on them. The contribution is not to state *that* new materialism applies to media but to look for *how* this might play out in their design. This search leads us to four main fields: media, design, performance, and craft. Media, performance, and craft each anchor their own main chapter, while design is interwoven across them. All four fields are interdisciplinary and notoriously hard to capture. Bridging across multiple disciplines we have to be selective. The result cannot be complete but hopes to invite other stories to detect and fill any gaps. To apply a clear perspective, the book establishes its own working definitions of performance, craft, and media. These might be limited in scope, but they are necessary to clarify the path taken through the masses of blending disciplines and blurring approaches. Each of the main sections will follow a similar approach: a focus on detailed example cases, a central section on theory, and a concluding example for key elements of the discussed theory in practice.

Each main section of this book opens and closes with individual projects that lay out the themes at work. These sections follow a close reading typical in material culture and the research concerning many of these examples included conversations with the artists and designers themselves. But to avoid a foregrounding of the human voice, these conversations are not included directly. There are no interviews included, but key elements of the artists are interwoven with the material work encounter. The role of the human maker shall remain balanced in relation to the role of the nonhuman components. It is one hope of this book to grant these projects, objects, and makers enough room to establish their own contributions and to provide them a voice through the methodology of the text. Nothing can replace the actual encounter with any one of them. Still, this is a book that tries to bring the material voices in media design forward. It tries to offer these pieces their own space to unfold.

In some cases, these examples are taken from well-established public projects; others are smaller case studies. Some of them are predigital; some, decidedly nondigital; some, hybrid. They reach from performance art to pottery to hybrid dance practices to personal fabrication, and they align more with the themes of their section than with any single platform or technology. Their selection does not follow a single media format or practice; nor was it limited to digital approaches. It was based on whether these projects have something to contribute to the central question of balance. This question is much older than digital media, which is why answers for this underlying core question cannot be found within a "digital only" frame.

Consequently, referenced works and examples reach across a wide range of designs from traditional craft to cornerstones of performance art to augmented reality pieces.

Following these opening case studies, the central parts of each section deal with critical reflections drawn from existing scholarly work. They form the theoretical backbone of each chapter. Following the practice-based nature of performance and craft, they look at processes as well as outcomes, whether these are objects or ephemeral performance happenings. Craft, performance, and interaction all share a fundamental dependency on action and direct entanglement. The goal is to unravel some of these entanglements through critical discussions largely centered on forms of production. They are interested in what is done, between whom, and what happens during that encounter. They do not provide historical reflections; nor do they tackle media philosophies or semiotics, but they trace a kind of evolutionary maelstrom of interdependent activities.

Admittedly, this includes a certain messiness. It is one preamble of this book that neat solutions cannot be the goal anymore. If we embrace Haraway's call to "stay with the trouble" (Haraway 2016), we must reject solutionism. The book does not offer a single neat framework that would solve the problems at hand. Instead, such claims for concluding resolves are seen as part of the very problem we are facing. Instead, this text might be better understood as a form of field diary, a journey of exploration based on encounters with central disciplines and ideas.

A principle shift, such as the material turn, requires a substantial reworking of our thinking. In such a reworking's wake, central works on media studies can turn into media philosophies. As John Durham Peters (2015, 12) argues: "Media studies is thus a form of philosophical anthropology, a meditation on the human condition, which also means a meditation on the nonhuman condition." This might be necessary, but it is not the turn that this book will take. Overall, Kember and Zylinska's exceptional *Life after New Media* might be the work that relates closest to this book. In it, Kember and Zylinska (2012) widen the role and activity of mediation onto the scope of human living itself. At the same time, they rightfully criticize the shortcomings of our scholarly scaffoldings. These shortcomings include the limitations of books, papers, or other accepted academic output. The media we produce are limited and obviously flawed, yet we are still operating within the outdated ecologies. Kember and Zylinska manage to leave this disciplinary straightjacket in their final chapter as they provide their conclusion in the form of a manifesto. This book attempts to turn their approach around. It introduces upfront a necessary ideal, vital media, and then it proceeds to discuss different aspects of it through the lenses of performance and craft. One might say that it starts with its manifesto to spend the rest of the text supporting it, reflecting on it, and positioning it in context.

A second difference between the present text and Kember and Zylinska's work is in the argumentation. If their book represents a creative manifest's encounter with media, this book turns to the logic of projects and things. Using close object readings in combination with critical theories, it moves in convergent and divergent phases that weave across examples, objects, and references to make a case for a different media design. It has great sympathy for practice, for the actual object, for the making, for the makers, for the tools in their hands, and for the materials on both ends of those tools. This verges on the anthropological, and over the past years, I had the opportunity to speak with designers, artists, and craftspeople about their practices. Those conversations steered the journey through questions regarding clay as much as 3D printing and hybrid designs. Only a few fragments of these conversations are ultimately included here, but the voices of these practitioners inform the text throughout. This might be the least visible methodology, but I remain most grateful for the patience and openness of these makers to share their experiences. The role of the maker and human contributor remains important throughout as the book seeks to position them into a form of applied media design.

In a mediatized society, media studies themselves are part of the problem and part of the challenge. As Kember and Zylinska (2012, xvii; emphasis in the original) ask: "Are our critiques not also forms of invention? Or, more broadly, can we think of a way of 'doing media studies' that is not just a form of 'media analysis' and that is simultaneously critical *and* creative?" This is a pressing question. If media are our ways to realize that we live in the Anthropocene, how can we conduct critical media studies *without* turning activist? Should we? For the time being, the established channels seem to hold. Books are printed; servers remain fed by power grids. But how to embody the inevitable adjustments in the critical discussions themselves? My own answer is certainly tainted by my work in the university, but in my experience, the most effective means for change are found in educational practice. This book hopes to stimulate such practice. It is not a summarizing textbook, and it rejects the promise of a single solution. It finds the idea of a particular invention or design to save the day to be a dangerous misconception. The argument strives not to end but to feed necessary debates, not to resolve but to emphasize the problems. This book is not a "how-to" guide but a media design exploration open for debate. It asks not only what the conditions are but also what we should do about them. That is indeed its task: to ask and invite its readers to keep asking. The target is to connect theory with hands-on material practices in that questioning. In that regard, it aspires to support the necessary educational steps to come.

Building Blocks

This book consists of four main sections, which move from a conceptual outline of the targeted vital media idea to two explorations across disciplines, to a concluding expansion of the original view. It first proposes a target, vital media, as the ideal that serves as a demanding guide for the remainder of the book. The following sections circle around different methods and challenges regarding ways to make these vital media happen. As the target was already introduced in the beginning, the argument ultimately does not arrive at a concluding resolution or second goal. Instead, it widens the focus from the individual that governed the original vital media concept outward to larger social groups in the final chapter.

We start with a view from space: the original earthrise image, taken by the *Lunar Orbiter 1* in 1966, opens the chapter "Mapping Vital Media." It introduces the concept of vital media in reference to the work of Wolfgang Welsch and Gilbert Simondon. This section settles the approach in material matters of media design and connects it to threads in new materialism, including Barad and Kember and Zylinska. But it does not replace the humancentric model with a materialcentric one. Instead, it deals with the human need for distinction through expression as much as humans' interconnectedness with the world they are embedded in. Humans emerge next to the materials and objects surrounding them. Each of these objects and materials features its own capabilities and corresponding actions. To highlight the human contributions, the argument focuses on cognition and the emerging need for expression. These stand in relation to the materials one encounters and within the cultural landscape that is being constructed. Vital media are defined as means of expression as well as means of physical changes. The "vital" connotation stands for the necessary balanced combination of the two. Arguing with Bennett, it emphasizes the dependency of cognizant and noncognizant actors in networks of activities. These networks are what make media happen. Notably, the actions within them have changed as objects have become increasingly technological. Here, Simondon's work assists in a turn to the technical object in the digital age and in mapping out the new milieu we are facing, which includes tangible interaction design, physical and ubiquitous computing, and hybrid objects.

Next, we turn to Joseph Beuys's *7000 Oaks* project, an urban social sculpture of city forestation. *7000 Oaks* introduces key themes of the "Performance Makers" chapter, which builds on basic principles of performance art and performance studies. The argument draws on key techniques from these fields to dive into the expressive side of vital media. It traces the expressive action in relation to the material and as a form of cultural production, referencing the work of Erika Fischer-Lichte. It builds on a definition of performance as *the action to produce expression with an intent*. This production-based approach focuses on the activity, the processes of shared production itself, which assemble into the building blocks of vital media making. Dragging materials, humans, and media onto a shared stage challenges each participant in its own ways. Each contributes its own elements to the unfolding media event but introducing materials as performance producers opens up new challenges. Such mixed forms of mediation complicate established qualities of performance, as a turn to Philip Auslander's discussion of "live-ness" shows. Combining forces of human and nonhuman actors also challenges the notion of "acting." This is explored through a turn to robots, who are discussed as active contributors without turning into human replacements. Balancing the forces of all participants, which include active code in video games as much as organic objects, ultimately leads to puppetry as a form of material performance. Dassia Posner, Claudia Orenstein, and John Bell are key references to support this point. The operation of a puppet serves as a culmination of the problems outlined in the chapter and as a powerful way for a material renegotiation (Bell 2014). The chapter closes with an examination of *Subway*, a project that reaches across many of the challenges outlined to that point by combining mediation with a distributed dance performance.

At the beginning of the next section, "Recentering," we find a cup crafted by Ehren Tool. Its markings as well as underlying practices introduce key themes of the chapter, which turns to craft and craft research. It avoids a nostalgic turn to a craft revival as it puts the "making" half of vital media production into context. This form of making is not detached from any expressive actions, and craft offers a rich perspective to map these interconnected activities. Drawing from Howard Risatti, it emphasizes craft-specific qualities, such as "need," as integral to the development of vital media through making. The working description sees craft as *cocreative material practices that are based on needs and further individuation*. It continues the balancing act that defines vital media by touching on materials, maker practices, and the objects that are produced through such hands-on engagement. It reframes the shared becoming and suggests the three steps of *encounter*, *exploration*, and *collaboration* as a way to frame it in practice. This practice has an educational undercurrent that brings us to critical making and critical improvisation, which leans on work by Matt Ratto, as well as Tim Ingold and Elizabeth Hallam. It should be obvious that craft cannot be reduced to a physical production only. Instead, this view of craft follows approaches in craft research that invite us to "think through craft" (Adamson 2007). The section closes with another example of ceramic work: Amit Zoran's hybrid reassemblage project that combines material agency with personal fabrication and design.

The last section, "Digital Folk," reaches beyond the individual's encounter with vital media and projects it onto a wider plane: that of folklife. It opens with Cat Mazza's *Nike Blanket Petition*, which foreshadows the theme of reclaiming a media ecology through community engagement. The argument combines folklore studies approaches by Richard Dorson and John Burrison with a look at particular practices among North Georgia folk potters. It distills three cornerstones for a digital folk idea, which focus on *lived material culture*, *variation within tradition*, and *community-based practices*. Manifestations of these cornerstones help in the final turn to *craftivism*—a term coined by Betsy Greer. Craft-based and digital-driven activism summarize these key points in selected examples, and vital media are not left standing as a theory trapped in ivory towers but opened up as cultural practice in everyday life.

A constant keyword throughout these pages is *balance*: the balance of ephemeral expression with seemingly dead objects, that of media activity with material agency, of making and performing, of communities with their environments. Balancing asks us to perform constant shifts, to keep moving, and to avoid rigid solutions. This book hopes to take the reader along for such a balancing act. It might not always move forward in a straight line, and it might move at irregular speeds, but along the way, it aims to offer moments that encourage us to realize that a certain equilibrium is possible once we put in the necessary work. These are the spaces that vital media lives in.

2 Mapping Vital Media

This section clarifies key terminology and connects media design with theories in new materialism. It develops the concept of vital media as a corrective and lays out the mission for the remainder of the book.

Media can get in their own way, as will be illustrated with the first earthrise picture taken by *Lunar Orbiter 1*. Looking back at the earth from space produced a new perspective and a new story. It also produced unique camera systems and a cosmic trash pile. The material conditions serve as a reminder of the limitations and connections that define the making of such an image and its consequences. Media design has to clarify how it takes on these webs of interdependencies. Building on new materialism, the idea of balance serves as a foundation to understanding the nature of vital media as shared activity networks. I will argue that the specific role of human contributors in these networks is to provide a form of cognitive willed expression. This is a defining element of media as well as of the human role within them.

The focus on these cognitive contributions is not a return to a mind-versus-matter debate. Following Wolfgang Welsch (2012a, 2012b), I contend cognition is instead situated in an evolutionary cocreation. The specific human quality of advanced cognition distinguishes humans from any other media partners, but it does not leave any room for false hierarchies. Instead, a discussion of the term *vital* identifies it as part of intra-action at work.

This view emphasizes the encounter between human and nonhuman, cognizant and noncognizant. It is the initial collaboration with a material object that defines the point of origin for vital media. But the nature of these objects has changed with the growth of hybrid materials and embedded technologies. Building on Gilbert Simondon's work I will discuss what role these hybrid objects play in the necessary balance of person and material. Following his concept of milieu, it becomes clear that this balance provides a much-needed alternative to the current unsustainable practices.

Finally, I will summarize key criteria of vital media to provide a basis as much as an initial scaffolding. Overall, this chapter introduces the call or mission brief for the rest of the book to follow.

Looking at Earth

On August 23, 1966, the *Lunar Orbiter 1* transmitted the first image of the earthrise, a view of our planet hovering above the surface of the moon. It was never meant to do that. The orbiter's main task was an exploration of the moon's surface to prepare for the forthcoming Apollo missions. The goal was to map possible landing sites, and this task drove the design of the *Lunar Orbiter 1* spacecraft. It had to transport a complicated photographic system, which affected the payload and overall specs of the craft (Byers 1977). But the conditions were not only set by these technical needs. *Lunar Orbiter 1* materialized in the hot phase of the space race, which was driven by technical as well as political agendas. The Soviet *Luna 9* craft had landed on the moon in early February. Its panoramic photo transmissions from the moon's surface were proof of the advances of Soviet power literally reaching new worlds and territories. The Soviet *Luna 10* had orbited the moon in April delivering a wealth of data. It also broadcasted parts of "The Internationale," the famous socialist hymn, back to Earth. This broadcast was timed to coincide with the Twenty-Third Congress of the Communist Party of the Soviet Union, where the transmission was celebrated as yet another victorious sign of their agenda. Political tensions on Earth reflected in the operations of spacecrafts racing to the moon, casting them in the light of ideological borderlines that were drawn. These tensions were part or *Lunar Orbiter 1*'s story as much as the mapping tasks and the technological implementation challenges.

Producing and delivering the image of the earthrise required substantial media technology, which itself reflected the historical and social changes. The *Lunar Orbiter 1* featured optics from a German company, Schneider (Kosofsky and Broome 1965), which had delivered lenses (including spy equipment) to Nazi Germany before the end of World War II. Images were captured by two onboard cameras using the Kodak Bimat process to deliver a fixed negative image. The Kodak Bimat process itself had been originally developed for US spy satellites. These images were scanned by a line scanner based on a CBS-developed film scanner (Kosofsky and Broome 1965) (see figure 2.1 for a breakdown of the process) and transmitted to radio stations on Earth, where "miracle-performing machines will convert the craft's messages into pictures" (MCA/Universal Pictures 1966), as a Universal Pictures newsreel explained.

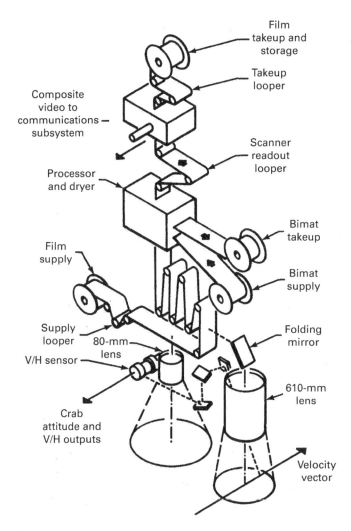

Figure 2.1
Diagram of the photographic system on board the *Lunar Orbiter 1* (Boeing 1967, 3).

Up in space, this equipment faced numerous challenges on its own. The Kodak Bimat process required certain temperature ranges; the special Kodak film equipment had a limited radiation resistance (a major sun flare might have threatened the whole mission); a special sensor was needed, the velocity-over-height sensor, which had to compensate for the orbital velocity of the spacecraft over the moon surface as it changed during its elliptical orbit. The closer the craft, the higher the velocity, and the bigger the danger for imprecise blurred images. Addressing these challenges was so central to the mission that delays in the production of the velocity-over-height (v/h) sensor and one of the lens shutters led to the overall delay of the spacecraft's launch (Byers 1977).

Because the mission was to map Apollo landing sites, the cameras were oriented perpendicular to the moon on the projected trajectory of the orbiter. They were fixed in the craft and could not be moved. However, once it became obvious that a shot of the earth was possible, NASA officials talked the manager of the mission, Robert Helberg, into temporarily repositioning the spacecraft to catch a glimpse of our planet from space, even though this required a complete reorientation of the whole craft. This risky maneuver might have threatened the original mission (Hansen 1995) but ultimately led to the first earthrise pictures taken August 23, 1966. It was a forced byproduct and not supporting the main mission goals. Still, the image of the earth rising over the moon immediately became a landmark achievement. It was prominently featured as a perfectly framed oversized proof of success at the press briefing on the achievements of the *Lunar Orbiter 1* mission (Hansen 1995, 345). The risky distraction had become a main attraction. Later versions of the earthrise included color images and higher-resolution pictures; some of those images were taken by humans orbiting the moon, some by humans standing on its surface. But the *Lunar Orbiter 1* was the first unit to deliver this image and initiated the powerful stories associated with it.

The metaphoric message of the image remains strong today. The perspective visually flattens hierarchies and questions divisions. The point of reference changes as we look over the moon's horizon and find ourselves looking back at our own planet Earth floating in space. One year later, Buckminster Fuller would coin the phrase "Spaceship Earth." Even amid the tensions of the Cold War, the image invited new interpretations and spawned new stories and reflections. The cameras and transmission systems on board *Lunar Orbiter 1* (see figure 2.1) provided the tools to contribute their means to such a new perspective. But within this ability is also the danger of error and abuse. For starters, the image seems to show the earth rising "over" the moon, while in fact it was flipped to achieve this effect. The original had the moon on the left and the earth appearing on its horizon to the right. This was set by the flight path of the spacecraft. Second, this flipped image was presented in a particular context. At the mission's press conference, a handful of white men unveiled it as a winning token in a military-industrial and geopolitical competition. It could easily be reduced to a conquest, an image to claim objective superiority, ownership. Images like these can be turned into monsters. Haraway (1998), for one, asks us to maintain the embodied view and to reject the "god trick" of a flawed objective and conquering gaze. The dangers and flaws in any such interpretation are part of the story. As a cultural object, the earthrise image and the stories that have been woven around it are certainly part of such fallacies. Any simplifying abuse of the picture for conquest, superiority, objectivity, unity, or any other agenda is flawed, and we should criticize the abuse of the image. But the image should not be reduced to the fallacies of a fake objectivity or a technologically disembodied feat. Instead, it is a picture of material agency first. It is a media piece produced in a particular way and fashion. In that regard, it relied on bodies and their actions. Many of them on the earth—others in orbit around the moon.

The first concept taken from the earthrise image as a material body emerging from a complex network of collaboration is that of self-infliction. Much has changed in space exploration since. New crafts reach ever farther, footprints have been left on the surface of the moon, and the earth itself is being circled by a growing barrage of monitoring satellites. In a reversal of fortunes, the production of one of the most iconic images of the space age is also a sign for the consequences brought about by such advances. As the mission came to an end, *Lunar Orbiter 1* was deliberately crashed on the moon and became part of an early space junkyard. The original image, which had been physically developed on board the craft, scanned, and transmitted to Earth, crashed with it. The same image, which was interpreted to speak about the volatility of a shared planet Earth and a concept of community on it, became part of a story about space debris and human pollution reaching new heights. It might inspire planetary unity, but it turned into one of the first trash piles left on another heavenly body. There are various indicators used to track the beginning of the Anthropocene, but leaving traces and pollution beyond the earth's own sphere offers itself as yet another reminder—a reminder also of the fact that ever bigger amounts of space debris have amassed in the orbit of the earth ever since. Debris circles the planet, crashes into satellites and spacecraft, and endangers future space exploration, as well as media and monitoring satellites. As spacecrafts reach out, they threaten our very capabilities of space exploration. Showing the fragile reality of the earth's atmosphere that protects us from the void, earthrise is not only an image of that vulnerable state; it itself is part of the trash piling against it. A first lesson, thus, is to accept the notion of media's materiality within and in the making of the Anthropocene. Today, NASA has a dedicated unit tracking space debris, making it all too clear that media design must be aware of its own trash piles.

The second concept is technical and material estrangement. In 1966, the mission not only worried about sensors and sun flares but also faced challenges on Earth. Data transmitted from *Lunar Orbiter 1* was stored in analog form on magnet tapes and was initially archived until 1986. They might have been the latest "miracle-performing machines" in 1966, but those miracles aged fast. When the tapes were returned to the Jet Propulsion Laboratory, the original data were almost lost—as were the tape machines, the Ampex FR-900s, that were capable to read them. The Ampex FR-900s had become antiques, and the instructions on how to operate them were missing. It took a dedicated project, the Lunar Orbiter Image Recovery Project, to find the necessary tape machines, assemble the knowledge and expertise to operate them, and reanalyze the original tapes. A team of dedicated enthusiasts had to actively counter the digital amnesia that had already set in. They rescued the tapes, reassembled and maintained the machines, read and reanalyzed the data, and ultimately delivered a higher-resolution image than the original. Only then did the image seen in figure 2.2 come into being. The image dates to 2008 and uses more recent data reprocessing to present a higher-resolution version than the original presented at the press conference in 1966.

Technologically, the miraculous computing power of the past has been surpassed manifold, and access to it has spread widely. That does not mean that the understanding of it among its operators has grown in equal measures. The newsreels of 1966 tout "finesse of computer legerdemain beyond the power of most of us to comprehend" (MCA/Universal Pictures 1966). The thrill of an expanding visual universe was still invoking curiosity and wonder. The computational power has multiplied in today's camera technologies, but the comprehension and curiosity of its operators seem to detach. Today's audiences are users, not puzzled by but fully expectant of ever-improving visuals delivered by their cell phone cameras. Even if they would bother to learn about optics or light sensors, why bother if their devices will be routinely replaced by the next generation? In too many

Figure 2.2
Image 1102_H2 originally taken August 23, 1966, recovered and reprocessed by the Lunar Orbiter Image Recovery Project (2008).

ways, technical advances and their speed have not brought us closer to a product but removed us further from the underlying operations. At the same time, the sense of wonder disappeared. Today's photo equipment easily outperforms that of *Lunar Orbiter 1* and is much more accessible, but we also disconnect further and further from the underlying material culture. What used to inspire awe and spark a national interest in the space race in 1966 has become a muted alien mystery, sometimes celebrated in advertising campaigns but of limited appeal beyond its functionality.

The challenge that technology posts to humankind has been part of different strands of scholarly work at least since the Industrial Revolution. In the digital age, this question of the technological has become ever more difficult to answer. *Production* of media has shifted to the idea of *use* of digital media. Material realities remain, but we have detached from the foundations of media technology tools, and this detachment is rapidly building up to a growing mantle of ignorance and disinterest. This affects the images and myths we live by. It is a danger zone that threatens shared growth much like the layer of space junk threatens space explorations.

If we do not understand the visual electronic scanning systems of the *Lunar Orbiter*, the metamedium of the computer adds yet another layer of abstraction. As Albert Borgmann (1999, 5) polemicizes: "Whatever is touched by information technology detaches itself from its foundation and retains a bond to its origin that is no more substantial than the Hope diamond's tie to the mine where it was found." Numerous scholars have already turned a critical eye on the promises of the digital age beyond the thrills of the initial discoveries and successes. Facing this danger, the goal here is not to complain about this detachment. It is not necessary to tackle the digital fallacy once more. Instead, we need to open up opportunities for reconnecting. To that end, it is necessary to clarify how to define media and their operations first, to recognize what it is that we need to reconnect to.

Media and Their Problems

The metamedium of the computer might have detached us from the knowledge needed to run an Ampex FR-900 tape machine, but it has added many of its own qualities as it has transcended into everyday life and materials. Digital qualities have become part of material media, and their ubiquity presents a massive opportunity. *Lunar Orbiter 1* was a rare object, detached and inaccessible for most yet inspiring for many. Its mechanized view back at Earth was unique and posed a new challenge. Today's applications allow a look at more distant as well as ever-closer components through the lens of virtual and augmented reality apps. But the danger remains that we fail to connect these viewings to our local, social, and historical ties, that we remain disembodied strangers in the depicted universe. This is the danger of realizing digital media as devices only and not as situated "things" (Borgmann 1984). If our looking is based on such a disconnect, then our perspectives are in danger to be tainted and incomplete.

Today's favored "look back" is not the earthrise, but it might be the selfie. We turn the gaze on our faces and inward into our bodies through smart sensors. We are being analyzed by the new miracle machines of modern medicine. MRI scanners, smart watches, and health apps are a given in today's mediascapes. They remain distant and strangely uninspiring, but they deliver the flow of information we live by in a digital "god's eye" back at ourselves. This disconnect requires us to reexamine their media designs. It asks us to trace the forces that drive ideas like the internet of things, ubiquitous computing, global internet access, social media, and their consequences. What the earthrise image calls for is not only a new understanding of media ecologies in the age of the Anthropocene but also a new turn in the ongoing development of our relationship with the material world and its agencies concerning media.

The story of the earthrise image is one of unfinished business, a chain story of objects, people, and changing circumstances. What is our role as humans entangled in these chains? This question motivated much of this book, but to set the stage, another challenge must first be addressed. In complex interlinked conditions like these, what is the difference between a mere object and media?

Matter and materials act. Plants, stones, animals, particles, and humans work in intra-actions that always intertwine and interlock. Material agency coconstructs the way we relate to each other as well as all other worldly conditions. To move forward, we need to distinguish the nature of media in these construction processes from the activities of other phenomena. In one way or another, media have too often been presented as centered on the human. Historically, their analyses focused on operations aimed at conveying something from one human source to another. In this understanding, a book would tell the story developed by a writer, a podcast broadcast the opinions of its makers, a film the vision of the director. Poststructuralism countered such authorial exclusivity, but the leftovers still echo in the way we see media and technologies today. In this view, different communication formats vary in their attributes and provisions. Television provides certain opportunities that differ from those available in newspapers or video games. It is up to the human to control, shape, deploy, read, and interact with them effectively. Such media properties are distinct mainly in the way humans speak through them. The idea of a medium as a "channel" changes into media as shaping forces, as "being the message." Here, media operate as rhetorical tools, as "any extension of ourselves" (McLuhan 1997, 7). But if we accept that materials are acting themselves, then we cannot claim that these extensions belong to us. The more we include new materialism into media studies, the more any hierarchical organization or ownership disintegrates.

It is no surprise, then, that media studies have turned toward nonhuman agencies and qualities to include natural and biological forces. The human body itself, formerly the centerpiece, becomes a complex "network" (Peters 2015, 6), and these networks operate at ever expanding scopes. In John Peters's case, they literally reach from the human body to the clouds above. Grant Bollmer (2019, 13) emphasizes the interdependencies further: "We do not communicate *through* language, or *through* media. Rather, *we are spoken by language. We understand ourselves* in *media*. Our knowledge of the world does not happen apart from the material forms that organize and shape the limits of relation, knowledge, perception, and memory." Any argument about media is itself embedded in such a material world formation. We are but one component, contributing our specific qualities to the unfolding activities. Stepping even further away from the focus on humans, other media scholars look at media beyond human communication (Galloway, Thacker, and Wark 2014). Given the original question—what is the role of humans entangled in material media chains?—this book's argument has to remain inclusive of the human role and draws the line in distinction to a posthuman approach. Instead, it follows Sean Cubitt's (2017, 4) view that "mediation is the primal connectivity shared by human and nonhuman worlds."

This brief overview does not attempt a history of media studies to illustrate the wide-reaching debates on media definitions and their undoings. Thankfully, other scholars have already traced this material turn in more detail. Particularly Bollmer (2019) lays out a most welcome introduction to materialist media theory and its development. Applying such material perspectives, Kember and Zylinska (2012) turn to the forces of mediation, which support a "lifeness of media" that is ever emergent. Following these scholars, the argument here will accept the formational role of materials. We will spend only little time tracing it back to the theorists of new materialism while trying to apply it moving forward. The goal is to explore what these networks and process-based conditions mean for media design. With that, we turn away from categories of media and toward the agency involved during mediation instead. This outgrows the need to define the differences between channels or formats. Dealing with mediation, we move ever closer to interdependent practices and to the actions of all partners involved. What defines media and the following discussions is a changing concept of agency and shared production.

Brought to its conclusion, the argument for material agency emphasizes that the action itself is always interdependent, and objects or phenomena only emerge through it. As Barad (1999, 826) argues, "agency is a matter of intra-acting: it is an enactment, not something that someone or something has." Barad's "intra-acting" has rightfully become a cornerstone of new materialism. Humans contribute to countless intra-actions: they eat, digest, breathe, move, build, and destroy, all the time in collaboration with the world they are part of and being shaped by. Most of these activities deal with making their existence in the world and thus are more focused on process than finalization. Action trumps fixation, but agency in this action is distributed, as Ingold (2013, 97) argues: "Humans do not *possess* agency; nor, for that matter, do non-humans. They are rather possessed by action." The question is not "who does what" but "what goes on in-between." These actions are forms of coproduction just like those of other critters and beings around us. They are on par with clay forming below and clouds assembling above. If intra-action is so central and the world comes into being through a shared collaborative effort, then it is a thorny problem to distinguish media from other activities. If all things have agency and all actions are interdependent, what is the particular activity or agency involved in media?

Media action becomes a form of "coemergence in which we shall also take part and through which we shall also become 'something else'" (Kember and Zylinska 2012, 186). Mediations, just like other actions, are forms of creative correspondence through which we change and emerge. The partners rub on and off each other to create the activity and consolidate themselves within it. The conditions for these encounters have turned ever more complex with the arrival of digital networks and the countless digital objects available today. These digital qualities add yet another layer of relation as the discussion turns further into systems, their organization, and orders of interrelations (Hui 2016). If that is the case, why use the term *media* at all?

The reason is largely the fallacy of humans as actors within these networks. *Media* as a term still make sense for two reasons. First, media production requires cognitive activity—and this distinguishes it from activity that does not. Second, there are differences of cognitive contributions—and this distinguishes media productions from each other even if other factors remain the same.

Media are active networks of entangled activity, constantly in a state of production. They are what Barad (2003) calls "apparatuses." But they are of an own kind. What differentiates media activity from countless other collaborative activities is the inclusion of a complex cognitive factor. Unless one dissects a thought to pure neural activity and disregards the result as an own entity, media activities cannot be described purely by physical forces. They are distinct from other activities because they depend on human contributions in the form of willed expression. These specific contributions depend on our cognitive abilities and shared needs that are still part of this world but that have distinct qualities.

Media remain shared emergent networks of activity—just as any other action. But at the same time, they are of a particular kind. They are children of cognition. The more complex the cognitive operations, the easier the activity qualifies as media. This definition does not reject cognition in other beings, as it has been discussed from plants to animals to cellular structures to shared cognition between entities (e.g., Hayles 2017). But it sees humans as particularly affected and thus tries to speak to them to adjust and evolve. Humans feature one of the most active cognitive systems we know; theirs is the biggest responsibility to realize it. As intriguing as these systems might be, their contributions are relational and always only operate within the wider network. They are a defining element but not the dominating one, much less the originating one. Human contribution does not begin or conclude. It does not close media production, which can never reach fulfillment. It is part of the networked process. That is why media and mediation are not owned by any single entity. They do not work *for* anything or anyone. They do not operate "by their own" or "extend" any central being. They morph out of and return to other actions as they emerge as entangled webs of interdependent actions. Humans are never the dominant factor in these moments of emergence, but they are tied up in these forming networks. These networks unfold via active contributions of all partners that, in return, shape these partners in the encounter. There is no center but a collaboration. Indeed, it is altogether misleading to argue that they "are" as they remain always trapped in a state of "becoming." Humans and all surrounding matter are part of this intertwined web of activity. The scientists and mission operators for the *Lunar Orbiter 1* were part of a complex network that also included multiple cameras, data storage technology, and specialized sensors. This network continues forty years later as seen in the image recapturing of the original data. Only through such a network can the earthrise picture be understood.

Countless intraoperations create objects-within-objects networks. Peters's concept of the human body as a network is one example of that. The operations of a 1960s spaceship are another case. At times, these networks become media—as seen in the earthrise example. At other times, they do not. Either way, the identity of any involved element within these networks can only be described in relation. The task of the human is not to claim ownership or superiority but to foster relations. Haraway (2016, 1) calls us to be "truly present" "as mortal critters entwined in myriad unfinished configurations of places, times, matters, meanings." Our presence can manifest in the countless actions that we are capable of but only within interconnected configurations. The individuality of any element stands only in relation to the ongoings in between the many actors. This entangled state of agencies relates to Kember and Zylinska's (2012, 57) notion of "critical attention" as it transcends human-centered intentionality. Humans lose their claim to single authorship, but they do not dissolve into insignificance.

Actions are nonhierarchical and interdependent. But that does not make them all the same. Activities differ based on the capabilities of certain nodes or subsystems. Evolution has diversified elements over time and thus allowed novel forms of activity to emerge. This includes photosynthesis, mineral formation, or cell division. It includes the pottery wheel, the printing press, Kodak's Bimat photographic process, and the computer. Media and technology blur for Kember and Zylinska (2012, 18) as they argue that humans have always been technological beings and thus also "always been mediated." They introduce a close neighbor to this book's title when they speak of the *vitality of media*, "the *lifeness* of media—that is, the possibility of the emergence of forms always new, or its potentiality to generate unprecedented connections and unexpected events" (2012, xvii). This is close to the vital media idea that should emerge from this chapter. But here, we differentiate between those events that emerge through activity excluding and those that include cognitive activity in the form of willed expression.

Media are emerging networks of collaborative activity that include human contributions through cognition as much as material operations. Because of the contribution from cognitive participants, media are unlike other activities and action networks. Just like other action networks, media networks emerge in between their participants. They are relational, and through constant conversation between their parts, they bring forth not only a particular network but also its individual components.

The question remains how to renegotiate our position within them. If humans remain in a particular position, yet the activities distribute further than has been noted before, how do we design for these human specifics? How to balance the needs and connections between the partners and find space for the human contributor? These questions do not attempt to put the human figure back into any center, but to answer these questions, we need to clarify the human's particular cognitive contributions.

Cognition and Consequences

Humans are cognitive beings who need to exercise this activity as part of their emergence, to construct the network that is their self. Yet it is important to keep in mind that while cognition distinguishes humans from noncognizant materials, it does not separate them from them. Cognition is an embodied operation, and thus it always remains embedded. This concept of embodied cognition is not new and has emerged as its own field over time. It has been discussed in many disciplines, including design (Abrahamson and Lindgren 2014), media studies, and performance and craft (Groth 2017). One related approach is N. Katherine Hayles's (2017) work on how nonconscious cognition affects the coemergence of human and other biological entities next to technical ones. Hayles sees cognition tied to systems, cognitive assemblages, that can feature distributed sensors, actuators, and thinking units. It is through the combination of the participants that cognition operates in a shared practice. The focus of Hayles's work is on an extension of cognition itself, particularly into the technical domain. It also speaks to the shift away from a human-centered model: "The emphasis on nonconscious cognition participates in the central thrust of decentering the human, both because it recognizes another agent in addition to consciousness/unconsciousness in cognitive processes, and because it provides a bridge between human, animal, and technical cognitions, locating them on a continuum rather than understanding them as qualitatively different capacities" (Hayles 2017, 67). Relating to Elizabeth Grosz work, Hayles supports the decentering of Grosz's biological view that avoids any definition of a new hierarchy. But avoiding new masters does not mean that subjects themselves need to be erased in the assemblage. They can still operate in combination with their partners, different, not superior, but still identifiable (Hayles 2017, 77). As will become clear, these points overlap with the argument for vital media. But whereas the more specific (and far more detailed) discussion on cognition in Hayles's work emphasizes the dispersed nature of cognitive assemblages, the focus of this chapter is on the combination of cognitive and noncognitive activities. It traces these interdependencies back to evolutionary relations, and it emphasizes two points: the necessary embeddedness of cognition in the world and the continuous emergent powers of evolution. Neither stands in opposition to such scholars as Hayles, but they provide a different direction to build on.

The philosopher Wolfgang Welsch approaches the relationships between humans and their surrounding world based on an evolutionary principle. This principle acknowledges the agency of materials and objects but puts it into an emerging bond. Agency evolves and grows to eventually build sensing mechanisms, such as eyes or ears. Things that have the ability to sense are then driven by their new capabilities to develop even more particularities of cognition. Sensing and acting systems can be found in countless varieties. Their cognitive abilities evolve further over different phases and include eventually the human kind, as one part of this evolution. Just as any other part, the human system features its own qualities and conditions that distinguish it from others. However, it is important to realize that these cognitive capabilities do not take humans out of the world. An increase of brain capacity or the fine-tuning of sensory equipment are evolutionary processes; they are not detached by some kind of human-centered perception but emerge from millennia of biological material evolution (Welsch 2012b). Forming eyeballs and visual nerves are developing systems in themselves and part of the need to see. They are part of the developing complex networks we have already met in the spread of agency to materials before. Developing such ever-increasing forms and combinations of cognitive abilities stands in direct response to the world that is perceived. The ability of the paramecium to detect the nutritional value of sugar, the ability of a hawk to see prey, the ability of dolphins to hear each other's voices are all directly connected to their being in the world—in a pond, in the air, or in the ocean. The cognitive abilities of humans are doing the same, never detaching them from their surroundings but emerging from an encounter with them. Consequently, *Homo sapiens* is only one piece in an evolutionary puzzle that unfolds over time, space, and form. "Our worldliness extends even to our highest activities, to our acts of knowledge. Hence the human is, from bottom to top, a world-connected being. The adequate concept of the human is *Homo mundanus*" (Welsch 2012a, 11; emphasis in the original).

The corresponding argument in second-order cybernetics, which has its own quarrels with cognition and developing complex systems, traces neurophysiological activity from basic receptor cells to more complex organisms to describe "recursive computation" as an ongoing reflection. This ongoing reflection should be understood as "the underlying principle of all cognitive processes—even of life itself" (Foerster 2002, 216). The hawk's eyes and their complex systems of translating light into information are only one part of a gradually optimizing cognitive system. "The nervous system is organized (or organizes itself) so that it computes a stable reality" (Foerster 2002, 225).

Cybernetics is concerned with systems, how they compute and how they affect each other. A defining element of second-order cybernetics is the realization that the observer of any system is not a detached onlooker but intimately connected to the observed, and "what is considered is not the observed (as in the classical paradigm), but the observing system. The aim of attaining traditional objectivity is either abandoned/passed over, or what objectivity is and how we might obtain (and value) it is reconsidered" (Glanville 2002, 3). Humans, hawks, or dolphins do not somehow step sideways into an abstracted form to build models of the world they observe. The logics of their neural processes and their realization through some form of cognition are integral parts of the system, which always also includes the observer.

Emerging cognitive abilities and noncognitive ones collaborate to bring each other forth. Because of the ability to see a mouse in the distance, the hawk can hunt it and affect the existence of other objects as the bird kills the prey. The mouse might have inferior eyesight, but it can smell food and affect a crop. The same atmospheric conditions that allow the hawk to fly and the mouse to smell also support the crop with water and light. "Support" might be the wrong word here—the mere existence of all these partners is interdependent on their shared activities.

New materialism does not only see the autonomy of the cognitive partners as they organize, learn, and optimize. It is not only about the ability to "make a distinction," as second-order cybernetics would call us to do, but it also recognizes all the other forces without necessarily framing them as systems of control. In an interconnected evolutionary view of relationships between beings, cognition is never the only force at work. The way water operates with clay, and clay is perceived by the skin, as the potter handles it to build a vessel while having a specific design in mind—all of these are equally ranked forces. They form a moment of balanced construction when the system is circularly feeding into and off each element within it.

That is why cognition is never detached from the proverbial primordial soup. Humans remain part of that soup as it has formed over the ages *because* of their cognitive activity. The origin and connectedness never detached from the most basic biological and pre-biological evolutionary processes. This is a version of the human that is entirely of this world because of their unique abilities. The role of the human of this world is critical-reflective as well as always connected. As humans produce critical reflection, self-recognition, and individuation, they are contributing to the world they are part of. They continue a long line of evolutionary steps that connect to their surrounding past and present. In that way, cognition and critical thought are not a differentiating spark detaching humans from their surroundings but the realization and development of their connectedness to the shared existence with the world they constantly emerge from. Practicing our part as humans is our contribution to the shared emergence. The chemicals needed for the Bimat process used by Kodak in the *Lunar Orbiter 1*, the evolutionary development of the hawk's eye, and the cognitive abilities of you as you read these signs are part of this shared history. They stand next to the activity of the ink as it interacts with the paper, the wind streams as they carry the hawk, and the pile of broken pieces left of the *Lunar Orbiter* as they slowly decay on the surface of the moon.

Welsch (2012b, 137) goes one step further and argues that through the evolutionary development of beings that act cognitively, the world constructed agents of her own cognizance ("Erkenntnis"), and "in our recognition, the world apprehends itself" (2012a, 11). Cognitive and sensing abilities are seen as components developed by the world and geared toward its own ongoing development.

In a mirror argument, Humberto Maturana and Francisco Varela (1992, 27) trace cognitive development as an always embedded but increasingly complex biological continuation where "bringing forth a world is the burning issue of knowledge." Knowledge, social life, cognition, and any other action are part of a world that these activities bring forward. Their focus is on knowledge in a cybernetic system where doing cognition is world creating. In contrast, Welsch argues for a form of world self-apprehension. To build the case for vital media, we do not have to subscribe to either the goal of "knowledge" nor "apprehension." If there is a reason for the evolutionary process, vital media are not it. They hopefully allow reflection but certainly do not contain it. Vital media as an ideal do not try to capture or frame a system, nor do they settle its purpose. But the focus on the biological embeddedness of cognition, which is central to Welsch as much as Maturana and Varela, certainly applies to vital media. It is one foundation for the becoming, for the processes of expression as well as material construction. It values the continuous coproduction of the environment these processes are always part of.

On the road to define media we had to recognize that objects, materials, and humans are active interdependent operators with their own agencies. All these actors are interconnected and are equally valuable as they form each other in the collaborative action network. But they are not the same. Operations differ in quality. They differ along demarcation lines, such as biological, chemical, or cognitive activities. Such differentiation does not suggest any stable hierarchy between the emerging relations, but it clarifies and emphasizes their interconnectedness and dialogue. Correspondence and coemergence thrive in differences. As actions gain complexity, from formation of elements to cells to increasingly complex organisms, the qualities of actions change, falter, accumulate. Cognition is one of those accumulations. Because it is part of an evolutionary process, it most certainly is not an exclusive property of humans. Haraway (2016), for one, elaborates on the cross-species kinship with her dog Cayenne, as one example. As noted, Hayles (2017) expands cognition into wide assemblages of collaborators, from plants to humans to computers. The evolutionary perspective presented here does not argue against the contributions of dogs or plants nor against that of minerals or radiation. Cognition is not reduced to a form of human thinking but is put in relation to nonthinking partners in an ever-shared emergence.

It was already established that all actions are intra-actions and that media are networks of such intra-actions that feature such cognitive contributions. These cognitive contributions are distinct from other beings. Terms such as *intent* or *will* are contested in the cognitive sciences, but to differentiate the form of human contribution, this argument will use the notion of *willed expression*. It is this willed expression that the human cognizant adds to the overall network production. How do we recognize the interdependencies of these media network activities? How do they connect to the needs and operations of the equally participating critters, minerals, material agents? This leads to the concept of vital in vital media: a quality of balance that is necessary for all partners involved.

Vital

Cognitive contributions go both ways. They are our contribution to the world we are part of, and they are the self-realization of our potential as cognitive systems. Media are active networks that contribute to the pool of actions constructing this realization of cognition through expression and things through material construction—whereby both remain interdependent. Balancing the cognitive with the noncognizant marks the emergence of the vital component. Vital encompasses the growth of both, the cognitive performance of the *Homo sapiens* and the contributions of noncognizant materials. Vital is a quality that supports the emergence of the individual partner and their abilities as part of a coevolutionary process of material development. Cultural constructs may grow out of these encounters, but we will not focus on these high-level traces and instead remain centered on the production activities themselves.

Nicolas Bourriaud (2002, 22) suggests the term "relational art" as an arena for encounters where the artist focalizes parts of these encounters and influences an unfolding collaboration: "Through it, the artist embarks upon a dialogue. The artistic practice thus resides in the invention of relations between consciousness. Each particular artwork is a proposal to live in a shared world, and the work of every artist is a bundle of relations with the world, giving rise to other relations, and so on and so forth, ad infinitum." Bourriaud centers on the social assembly in the turn to this art form in order "to live in a shared world." It is most inviting to adopt it as advice for the media designer, too. But "life" is a contested quality that might get in the way when we want to define vital.

The idea of vitalism builds on the differentiation of the living and the nonliving. Vitalism was constructed around the idea of a special quality that appeared to be available only to the living. This divisive quality, the spark, could all too easily be abused to distinguish humans above all. To adjust the term to the flatter hierarchies of a collaborative and cocreative universe, vital will need clarification in the new order of plentitude.

Bergson casts a wide net in his investigation of human life, human knowledge, and intellect. He provides one possible answer to the question of what vital means with his concept of the "elan vital," which is made possible by the "reservoir of indetermination" (Bergson [1911] 1998, 126) of a neural system. To create, this force works with its surrounding matter;

it "seizes upon this matter, which is necessity itself, and strives to introduce into it the largest possible amount of indetermination and liberty" (249). For Bergson, the way of this capture is defined by a range of forces leading up to the "original impetus." Many of the conditions of his argument—originally formulated in 1907—have changed since, but this encounter with matter through *active interference* as a defining moment for the vital impetus remains inspiring. Using Bergson as one of their key references, Kember and Zylinska look at the vitality of media and a "becoming-with-media" (204) based on the premise that "we are—physically and hence ontologically—part of that technological environment, and it makes no more sense to talk of *us* using *it*, than it does of *it* using *us*" (Kember and Zylinska 2012, 13; emphasis in the original). They argue that "mediation can be seen as another term for 'life,' for being-in and emerging-with the world" (23). Their ambitious project centers on the concept of "critical attention," which "transcends human-centered intentionality" (186). Kember and Zylinska adapt this concept from Foucault and describe it through Barad's widened concept of agency that we have already encountered above.

In comparison, this book takes a turn from Bergson's discussion of the elan vital, where it reaches into fields far beyond expressive media constructions to look at life itself. The argument here does not follow life itself, but it leans on Bergson's concept of the evolutionary nature of action as a *creative act*. To be more precise: it is a cocreative act. Collaboration takes center stage. From here, we can ask how the balanced collaboration affects media design specifically. To answer this, we will not define anew what life might mean. Instead, we turn to the processes themselves and lean back on Welsch's evolutionary approach. Just as Welsch argues that human evolution and its development of cognition remain steeped in the world, the idea of vital remains connected to an of-the-world evolutionary condition. Intentionality does not disconnect it from becoming with the material world. It is not a hierarchical separation of an "alive" subject over "nonliving" objects. The processes remain part of the shared evolutionary growth among all partners involved. *Homo mundanus* is only possible as a body-of-world, and it is a constant building site of ongoing coemergence. For better or worse, the development of human cognition is one of these emerging qualities. We are part of the building site that includes the development of critical reflection and individuation among many other things. Vital asks us to maintain that building site for all partners involved.

This building site, the collaboration of the world we bring forth together, is in peril because humans have distanced their contribution. This includes the often-stated differentiation of humans from nonhumans through our ability to master language and abstract representation. Along these lines, media could be misinterpreted as channels to feed such communication. This would literally silence all other elements at work. Thus, instead of turning to communication as a distinctive feature of humans, vital turns to collaboration with material. Language remains one of the critical components that humans contribute, but the critical reflection and the interoperating processes of vital are not centered on it. There might very well be a "cognitive circularity" (Maturana and Varela 1992, 244) that allows humans to use their tools to think about knowledge itself, but vital media do not center the human anymore. Supporting thinking about our own knowledge as something beyond the limitations of the world is not the task of the vital media ideal. Vital media do not subscribe to hierarchical constructions and instead rely on interdependencies. Thus, the main quality of vital is to further advance all partners involved. It requires us to differentiate between the individual actors, but it forbids us to elevate one over the other. Instead of vertical expansion—usually for the sake of humanities' exclusive role and its justification—this requires horizontal expansion across the field of collaborating and coconstructing partners. A great many forces—biological, mechanical, chemical—are at work here. They enrich but also complicate any analysis of these networks. Comparable shifts can be found in related approaches. Actor network theory (Latour 2005) has been especially successfully applied to media studies. But unlike actor network theory, the networks of new materialism are not connecting existent partners. Instead, the unfolding action creates these partners. We cannot presume preexisting points that might connect but instead are left with an ever-expanding bundle of connections that constantly build forward. They all support some form of expansion through evolutionary processes. Cognition, experience, and critical thinking are part of them, but so are gravity, cell division, and televisions. Vital is proposed as a corrective to make sure that these shared realizations emphasize this connectedness. Too often design criteria push human-, experience-, or user-centered designs, which might justify vertical hierarchies for the benefit of the human instead of the vital horizontal codependencies for the benefit of all.

We have lost our central spot as the only, or even dominant, active entity and we have to learn to live with it. Barad (2007, 33; emphasis in original) makes the case that all agents, human as much as nonhuman, construct each other through a form on entangled "intra-action," wherein *"agencies are only distinct in relation to their mutual entanglement; they don't exist as individual elements."* It builds on a larger concept of "aliveness" as suggested by Barad (2007), who sees vitality spreading into zones that had been barred from it. In her own take on the call for new agency, Bennett (2010, 3) gives "voice to a vitality intrinsic to materiality, in the process absolving matter from its long history of attachment to automatism or mechanism." Bennett's "thing power" describes the agency of stuff, not as observed objects that come into being through human perception but as self-standing things. A life, in Bennett's case, is in immanence, unfolding, never settling, building on Deleuze and Guattari's "matter-movement." It stands in contrast to a model that sees life dependent on anything nonmaterial. Artisans cannot claim any "formative power," but "in this strange, *vital* materialism, there is no point of pure stillness, no indivisible atom that is not itself aquiver with virtual force" (Bennett 2010, 57; emphasis in the original). Potentialities of this force are ever active in all things interweaving with each other. Building on this entanglement, we follow the idea of vitality forming in an in-between, but the goal of vital media is not to replace one center (the human) with another (the material). Acknowledging the importance of material agency also does not relieve the human of theirs. Bennett and Barad have their own differing readings of this material agency, but for vital media, their shared focus on relation is more important than the differences in their takes on materialism.

Ascribing things, networks, microbes, or any other object the power to make and act helps in the necessary rearrangement of the human and their role in the world, and it spreads the notion of "life" across a wider range of collaborators. Human "life" is of, among, and only in conversation with material agents. This opens up conflicts with what human activity can contribute to the unfolding conversation. On the one hand, humans have to humbly accept the flattened hierarchies; on the other, they have to responsibly realize their own distinct contributions. Humans might have to share the stage with all other beings and "thing power," but they still contribute their own distinct actions to the play.

So far, the argument has emphasized the seeping of material, noncognizant forces into the balancing act that is vital. It leveled out the hierarchies without equalizing all contributors. Now we turn to the human contributions. Such a focus on human contribution should be read entirely within the balancing efforts, not as an undue elevation.

To further humans as material and cognitive beings, vital requires expression. Expression is needed for individuation particularly between humans, who can reflect on themselves. This is not a challenge that many other agents on Spaceship Earth face. Networks and microbes act. They might very well be part of a shared form of cognition and feature sensing and computing mechanisms. But they do not act to consciously self-express. Humans are among those cognitive beings that have to create expression to further develop the operation of their cognition. As argued with Welsch, this construction is part of our contribution to the evolutionary trail. Individuation is not a detachment from the world but part of a person's realization of their human potential within it. When it comes to media—those action networks that include a willed expression—the expressive impetus originates in the cognitive partners. They require it, much like a TV set requires electricity or a fountain pen requires ink. We cannot solve this challenge through objects that are not concerned by this necessity, like computers, TV sets, or books. But solving this challenge we must, for each single person's sake and to renegotiate the material conversation. Balance, coemergence, and interdependency are the cornerstones of vital, and they require us to realize our own operations in this balancing act. Vital is the quality best suited to express the necessary balance and to address this repositioning.

Vital media are intradependent insofar as they realize differences between partners, but they see these differences only emergent and active when these partners work together. They do not exclude agency from any active object; nor do they set human agency over any other's. They are based on coexistence and a becoming with but also on instability and differences. They are defined by an act of balance. Any such balancing act requires instability and dynamic forces. Vital media cannot be resolved but only managed. *Homo mundanus* becomes an active partner in this balancing act through its contribution of expression. Willed construction is certainly not the only action of humans, but it is typical for the cognitive beings that we are. Vital media fully accept the important and active role of other components in the collaborative construction of any such expressive action. But only an entity that requires self-construction is capable of providing the action of expression into this collaborative pool. This contribution can transform the shared pool of actions into a media activity that harbors ontological powers. Vital media that emerge from cognition and material activity are construction processes, not mere reflectors of some other reality.

We arrive at an understanding of interdependencies wherein the emergence of a person is intrinsically linked to the emergence of the world they are of and vice versa. It is part of human nature to differentiate, to individualize, from the surrounding others and nonhumans. It is part of humanity's contribution to the world to realize this specific action within it. Cognitive activity and willed expression are defining human qualities and yet always in reference to the emergent whole. It is through them that we make ourselves actionable in relation to, not better than, other beings. Vital media are the ideal emergent action networks that form out of this.

This turn includes a step away from the dominance of media formats that remain human focused. It stands against a focus on social media, for example. Discussing media with a focus on interpersonal communication is crucial to develop them further. It remains part of their operations. But making it their focus might step over the initial moment of material construction and intra-action. Communication and cultural construction are essential, but these are second-order effects. The first line of activity deals with the individual encounter with the material world. This encounter consists of at least two contributions: a factual material change in the dealings with the world, an "act on inert matter," as Bergson would argue; and an act of expression driven by cognitive activity, an act in which the human constructs and "the world apprehends itself," as Welsch would argue. In combination, it is collaborative action as self-construction as part of world construction. Vital describes the necessary balance that keeps these networks operating—to keep them and their partners alive, if you will. As we define ourselves against other humans and other stuff, and as we contribute to the construction of the world, vital is a dialectic quality of that relationship. The human body is not approached as more valuable than that of *Lunar Orbiter 1* or an ant's or the moon. But our self-positioning toward these things and others puts us in critical relation to them, contributing to the construction of both in that process.

These sections have traced two strands: on the one hand, our cognitive abilities emerge as part of a material evolution history; on the other hand, these abilities only manifest in action through cocreative activity with other biological and nonbiological actors in a shared world-creation process. Vital is the quality of critically balancing these two for the further development of both. Media are the active networks unfolding to make this happen. Both combined call us to look further into two directions that shape these unfoldings. First, we need to look into the construction of self through the production of personal expression. Second, we need to look into material construction processes through material making. These are the scales that need to be combined and balanced.

So far, the encounters with the material world have been largely centered on physical and mechanical components. But the digital has infused our world with new opportunities and challenges. Any object of the digital age includes potential procedural logic beyond analog mechanical logic that adds new behaviors. Objects are not what they used to be. To make vital media applicable to hybrid forms of technologies, we need to look into the nature of these technical objects and the growing complications they introduced.

Simondon and the Technical Object

Making expression becomes even more complex when it includes hybrid materials or blended objects. A new layer of complexity is added to the already challenging tasks at hand. Material agency expands as formerly seemingly inanimate objects have sensors and digital components embedded. Those can change their behaviors in significant ways. A car might navigate differently based on embedded computer systems that track its position and relation to other cars; medical devices might automatically collect data about the patient they are attached to; ever smaller devices are embedded in ever more objects to form new networks in what adds up to big data. These networks add additional qualities, a digitally defined layer of agency on top of the layer of material agency that any of these objects (car, body, object) had in the first place. Both merge and challenge what it means to interact with these hybrid things. Even media objects that were created seemingly exclusively through handiwork are codefined by their construction in a digitally infused context. Objects finished before the invention of the microprocessor are being reframed, often enough also remediated, into a new context. William Morris was inspired by a bird stealing strawberries in his countryside home to create one of his iconic designs, *Strawberry Thief*, in 1883 (figure 2.3). It took Morris multiple years to optimize the printing process of the design before he finally managed to use a traditional indigo discharge printing process to achieve the desired results at his factories at Merton Abbey in Surrey (Davis 1995). That process made this pattern one of the most expensive ones available from Morris.

Figure 2.3
William Morris (1883) *Strawberry Thief* pattern print (V&A Museum no. T.586–1919).

Originally intended for such textiles as curtains, drapes, or covers, today that very design can be found on any surface imaginable thanks to digital imagery, printing, decals, and other forms of reproduction. It is printed on T-shirts as well as on milk cans. This changes the role of the original design from one of exclusivity to a widely accessible commodity of a pattern. It also changes the individual realization of each instance, as few (if any) of the reproductions follow the original printing process used by Morris. Thus, the results and their historic context will always differ from his original production. This is part of the logic of remediation, as introduced by Jay Bolter and Richard Grusin (1999): one media format asserts itself into another, and new interrelations emerge between them. The more the remediated reference changes in functionality and gestation process, the larger the shift. In digital media, this shift is still growing.

Whether it is through new functionality, distribution, or novel production techniques, through the digital we face the object in different ways. If the logic of Barad, Bennet, and Haraway introduced material agency, then the logic of physical computing, the internet of things, remediation, and material-centered design add a computational quality to the already activated matter. Mikael Wiberg (2018, 6) has argued for a "material-centered approach to interaction design" to investigate this material turn in computing, and he builds on the idea that "if computing is increasingly less about digital interfaces and representations, and increasingly more about objects and materiality, then interaction with and through computers is radically changing as well." Interaction design, in this view, has to turn to the underlying "raw materials" (2018, 10) and their properties. At the same time, the features of digital media hybridize these materials. The ubiquity of the digital adds new qualities to an object's agency, such as procedurality or numerical representation, qualities that have been discussed much earlier as foundational to digital media, for example, by Janet Murray (1997) and Lev Manovich (2001). They become technical machines or add a technical machine quality to objects.

Writing in the time of the first microchips and at the dawn of the cyber age, Gilbert Simondon approaches the encounter of human and object in his own way. A reading of some of his ideas helps to explore this dimension of hybrid objects. Simondon ([1958] 2017, 16) claims that we misinterpret machines in our culture. He argues against the idea of the machine as a "stranger" "inside which something human is locked up, misunderstood, materialized, enslaved, and yet which nevertheless remains human all the same." Instead, he sees machines as "mediators between man and nature" (15) and man as a form of organizing conductor. Nature—in Simondon's view—offers already optimized objects and processes. They were adjusted and perfected through millennia of evolutionary alterations. In comparison, machines are objects that are still on their way to possible optimization. They gradually adjust as they are being operated and improved. Their human engineers, conductors, and operators are integral parts of these adjustments. Man "has the function of being the permanent coordinator and inventor of the machines that surround him. He is *among* the machines that operate with him" (18; emphasis in the original). The human conductor's goal in this concert of machines, machine parts, and material operations is to make the object more concrete.

For Simondon, concreteness depends on the operational optimization and technical comprehension of the object at hand, its functions, and its subfunctions. Morris's yearlong attempts to optimize his dyes and printing techniques to ultimately deliver *Strawberry Thief* in the quality he had envisioned represent one such optimization. The development of a motor from an initial prototype to an optimized efficient engine is another. The optimization of the frame rate in a video game is a third example. Along the way to such improvements, iterative changes can range from ironing out detrimental side effects to the realization of completely new objects. Humans learned about the effects and properties of steam and built a steam engine. This knowledge allowed them to envision new objects and develop novel scientific plans that led, for example, to the wider availability of electricity. They can also optimize existing ones, such as improving the operation and efficiency of a steam engine itself.

Simondon ([1958] 2017, 38) argues that, through such optimization, function builds on function to an ever-improving collaborative whole, and the object "is no longer in conflict with itself." Quite literally, this can refer to friction, energy loss, efficiency, and procedural optimization. It is the dialogue between machine and human that propels the perfection of the technological realization. Objects become more and more "concrete" as they grow from their initial designs into improved versions. According to Simondon, "concretization gives the technical object an intermediate place between the natural object and scientific representation" (49). Because technical maturity is needed to reach this understanding, Simondon differentiates between a "childlike" and a "mature" human encounter with the technical object. He proposes a "minority" status to daily-use objects that might be explorable and useful to a childlike, somewhat naive encounter but cannot be altered or understood. Any advanced, or mature, encounter is dependent on a "majority" status, where expert human skills define the relation to that object. Between these two, we can sense a detachment. Simondon argues that humans had historically lived in unity with technical objects, but they lost this unity:

> The unity of technics did not persist; a genuine reversal took place, which repressed the ancient noble techniques (those of agriculture and animal husbandry) into the domain of the irrational, the non-cultural; the relation to the natural world was lost, and the technical object became an artificial object distancing man from the world. We are only now beginning to see the possibility of an encounter between the way of thinking inspired by a technics related to living beings and an artificialist way of thinking concerned with constructing automata. Mechanical technics were only truly able to attain majority status by becoming technics thought by the engineer, rather than remaining the technics of the craftsman; at the artisanal level, the concrete relation between the world and the technical object still exists; but the object thought by the engineer is an abstract technical object, unattached to the natural world. In order for culture

> to incorporate technical objects, one would have to discover an intermediary between the majority status and the minority status of technical objects. The condition of the disjunction between culture and technics resides in the disjunction that exists within the world of technics itself. In order to discover an adequate relation between man and the technical object, one would have to be able to discover a unity of the technical world, through a representation that would incorporate both that of the craftsman and that of the engineer. The representation of the craftsman is drowned in concreteness, engaged in material manipulation and sensible existence; it is dominated by its object; the representation of the engineer is one of domination; it turns the object into a bundle of measured relations, a product, a set of characteristics. (105)

Simondon maps this disjunction and repositions humans within it. He argues that humans need to find ways to construct—"mediate"—themselves in this world of technical objects: "Man no longer needs a universalizing liberation, but a mediation. The new magic will not be found in a direct expression of the individual power to act, assured by the knowledge that gives each gesture effective certainty, but in the rationalization of forces that situate man by giving him meaning within a human and natural ensemble" (119).

The problem for this realization is that the rationalization becomes ever more complex. Many technologies of the twentieth century remain beyond the grasp of any single individual. The field of cybernetics faced this charge of communication and control within increasingly complex networks (Wiener [1961] 2019). It traces networks of control and the effects of objects and systems on each other. But initially it also had problems positioning humans within that logic. It required a turn to a second-order cybernetics to include the observing human within these networks (Glanville 2002). Simondon developed his own argument in the late 1950s, at the dawn of cybernetics. He introduces his reading in reference to cinema and television but reaches forward to the digital age and points out that humans are alienated by isolation and the disparate information brought upon them by the digital turn. Navigating the detachment of technology and its agency from the human grasp and understanding has not become any easier since.

To achieve the proposed ensemble and humankind's role within it, Simondon ([1958] 2017, 1992) suggests the process of individuation. Humans shed their role as the increasingly detached tool bearers they have become through the ages. Instead, they reposition themselves and become empowered to reflect upon the technical structures that coshape them. Likewise, cybernetics provides new tools for humans to not only produce but also to reflect and organize. It establishes new opportunities to look back at one's role within any environment. It establishes a new dialogue, which is ever evolving. Because we are constantly empowered to gain such a new position, any finality of human's relation to the technical is broken up. Outcomes cannot be seen as complete or finalized anymore. The process, the "creative force" ([1958] 2017, 120), becomes more important than any intermediate state or result. This ongoing process itself is central. We will never be done with the encounters Simondon describes. We will always discover new components and layers. We have to keep on inventing ourselves. That is why a multiplicity of conditions, continuous becoming, and openness of material engagement are not confusing or disorienting. On the contrary, they are liberating and signs of the emerging creative powers at work. Realizing these powers leads to the process of individuation. It is necessarily a perpetual process that remains without conclusion, an eternal construction. It can neither consolidate; nor can we ever presuppose a completed state; instead, it is "life itself following the fundamental mode of becoming" (Simondon 1992, 305).

For Simondon (1992, 300), the "individual is to be understood as having a relative reality, occupying only a certain phase of the whole being in question." Individuals qualify by continuing in this relativity and by never stopping the process of relating. Another way to describe this is as *"conservation of being through becoming"* (301; emphasis in the original). This becoming is tightly connected to one's surroundings, time, and conditions, and the result is the entrance into the milieu, another key term for Simondon. In a shared milieu, human participants collaborate with technical objects.

On the one hand, individuals evolve in such a milieu closely interrelated to all their surrounding technical objects. They are embedded in it. On the other, the differentiation of individual versus their milieu is central to this process. Human conductors do not dissolve into the technological but emerge in relation to it through a process of constant becoming. They define themselves through connection to as well as through differentiation from the surroundings they are part of. They can never reach a stable equilibrium, and Simondon (1992, 304) discusses it as a "metastable system" that might be full of potential but where "neither form nor matter are sufficient. The true principle of individuation is mediation, which generally presumes the existence of the original duality of the orders of magnitude and the initial absence of interactive communication between them, followed by a subsequent communication between orders of magnitude and stabilization." The more we relate, the more we stabilize.

This optimization is informed over time by science and manifests in the improvements of the "concreteness" of the technical object that the human achieves through the application of knowledge. In the advanced stages of such concreteness, the technical object is compressed and optimized to a level of highest possible saturation from an initially largely intuited one. For example, optimizing materials in a particular object can lead to optimized stability, functionality, and production. These optimized objects are comprehended by technical engineering, design, and science. This applies to William Morris's experiments with indigo dyes as much as to the development of steam engines, the *Lunar Orbiter 1* spacecraft, or video games. In the same way, the encounter with the digital activity is an ongoing individuation of the human, an optimization of the technological object toward its concreteness, and the place where both "become" as growing partners in a shared enterprise.

Like any other matter needs to be balanced, digital components are never conclusive and always in need of further optimization—not only for their own sake but also for that of their human collaborators. Only in the dialogue, in the cooperation, and in mutual construction do they succeed. Only through this dialogue can the process of individuation unfold. This provides a powerful expansion of shared material encounters into the realms of technical objects and their changing operations.

Simondon talks about the mastery and control of humans over technical objects. Even though the creative powers are distributed, this does not directly map onto the flat hierarchies suggested by Barad, Bennett, or Haraway. Yet, in the construction of vital media, the open individuation is one building block adopted from Simondon. Humans need to continuously reinvent themselves through embedded encounters. But within that ongoing construction, they shift the "creative force" further into the idea of intra-action developed by Barad. Differentiating between minority and majority status might help to inform the encounter, but whether one is "drowned" by objects or "dominates" them is a matter of perception and less important in a vital media view. Mastery of technology is not a necessary end product, and the eternal encountering is much more important. Thus, as we adapt Simondon's key principles, such as milieu and individuation in the encounter with the technical objects, at the same time, we shift them into flattened hierarchies.

This short turn to some of Simondon's key concepts provides a clarification of the added challenge that digital technologies bring to the material encounter. It reframes a call for material-centered design for the technological. The focus is on the never-ending operational encounter and the individuation that ensues. The technical object cannot be the solution itself, and the disconnect with the technical object cannot be designed away. We cannot make it disappear with some ingenious tangible interface. But design might support work within that problematic zone and help human participants recognize and grow in relation to the object and its operation, differentiate from it and thus emerge in relation to it.

This does not simplify the media design at hand. It complicates a great many qualities that are often seen as central to successful design, such as "immediacy," "natural interaction," or "intuitive interfaces." But it also emphasizes the enormous value of properly designed encounters with those technical objects: we need those encounters to form ourselves as much as the objects and world components around us.

Vital Media Call

The ideal of vital media builds on (1) the evolutionary positioning of the human as embedded in the world because of their cognitive abilities and (2) the emergent role of object agency and its complications in the digital age. One way to visualize the interdependencies is the image of the web. Each contributor to the structure is essential to the whole as they help to bring the other partners into being within a shared whole. In such a web, no strand can stand alone, and it does not presume a particular shape or individual anchors upfront. Haraway (2016, 13) speaks of "string figures" to describe the interdependent relating, the world creating, between all critters and other beings: "Natures, cultures, subjects, and objects do not preexist their intertwined worldings." From this, Haraway develops the larger concept of the Chthulucene. In contrast, vital media do not expand but are more specific to particular kinds of these intertwining relations. They focus only on worldings that deal with forms of creative expression.

They form in the encounter between the human and increasingly technical objects and can trigger the individuation of humans as well as the concretization of nonhumans. They happen during a creative development unfolding between cognitive humans and material-technical objects.

Ultimately, these encounters cannot be resolved but only managed. This management is the role of the media designer. On this level, vital media key activities can be defined as

(1) the realization of the material agency (i.e., the inclusion and recognition of the material's forces),
(2) the realization of human cognitive abilities (i.e., the support of willed expression as a contribution to a shared emergence),
(3) the concretization of the object (i.e., optimization toward an ongoing emerging next state of material balance), and
(4) the individuation of the human (i.e., integration as ongoing emerging differentiation from surrounding objects in a shared milieu).

Individuation depends on a self-positioning of humans and requires them to constantly create their evolving position next to these materials. Creating this position and partaking in the concretization of the technical objects along the way is the creative construction of the shared milieu. A balanced milieu is the playground for vital media. On this playground, they are bound to share such qualities as the sustainability of all partners involved and a continued emerging exploration. These qualities provide a shared center of intra-action around which practices can gravitate, never standing still, always evolving. Or as Heinz von Foerster concludes for the construction of any reality in second-order cybernetics:

> What are the consequences of all this in ethics and aesthetics?
> The *ethical imperative*: Act always so as to increase the number of choices.
> The *aesthetical imperative*: If you desire to see, learn how to act. (2002, 225; emphasis in the original)

Practically, such construction happens through rich expressive as well as productive actions. Humans take their part in the making of the objects, selves, and construction of the shared milieu, but it is only one part of this construction work. For objects, concretization continuously aims to optimize and bring forth "more choices" of the object in its relationship with the surroundings. Design inspired by vital media assists humans in their individuation through continuous, nonconclusive creative action; it assists objects in their emergence through ongoing improvements utilizing human collaboration; it assists the emerging shared milieu through a nondestructive quality of balance. Vital media depend on sharing growth. They are expressive and productive and do not have to differentiate between the two, as they realize their transformative impact on human cognition and object matter alike.

A second visual metaphor for vital media would be a mobile. All partners in a functional mobile have to balance out, no matter how complex and varied a single weight is. All partners have to contribute their own weight/agency to support the balance as it might shift due to external forces or the addition of more weights. Such a balance inherently mirrors a sustainable approach. In the long run, a vital media milieu cannot be self-destructive because self-destructive tendencies would be precisely those qualities that would need to be optimized and balanced out.

If these are the qualities and defining processes in vital media, where can we find them? We have introduced vital media as an ideal, an optimized equilibrium. But any stable state is impossible in the underlying evolving evolutionary system. This is why vital media are an ideal. They strive for optimal balance but also depend at the same time on the balancing effort itself to come into being. Vital media operate through the very mechanisms that make them so elusive. They are not a form of media technology but an optimization effort for all of them. The target is not to define a specific media network, designate it as vital media, and argue for it in comparison to other systems. Instead, the goal is to establish vital media as an ideal to inspire constant improvements to as many media systems as possible. This establishes the conceptual manifest and aspirational ideal upfront. The remainder of the book chases this ideal and highlights pathways to support it.

3 Performance Makers

How do vital media support an individual's creative expression? To tackle this question, we turn to performance. Its first example, Joseph Beuys's *7000 Oaks*, connects human performance to that of stones, trees, and cities. These connections are tested in Eva & Franco Mattes's reenactment of the piece in the online world of *Second Life*. What remains of Beuys's original social sculpture in a world of online virtuality? And what changes? The goal is not to provide simplified answers but to open up the problem space of performative action ahead of us.

With this in mind, we briefly review why performance studies is a useful lens for digital media. Performance offers a critical lens and practice to hold to vital media. It is a useful complication that allows us to keep the focus on process and the production of expression. To sharpen this production-based view, a performance is defined as the action to produce expression with an intent. Performing is production first—in material encounters and willed expression. How do individual cognitive contribution and material agency cocreate a performative media moment? This leads us to a building block of vital media: the shared bringing forth of a performative moment.

Once this collaboration is clear, we look into the role of the partners involved, including humans, objects, and media. They coconstitute each other in their encounters but contribute different qualities to the process. Mediation in digital media has challenged concepts of liveness and presence and added a new range to the available actions. But they still require human interaction to become performative. Shaping this collaboration is the task for vital media making.

Finally, we discuss how this function can be applied in more detail. This leads us to puppetry and material performance. They provide practical examples for the merger of activity and expression through a collaboration of cognizant and noncognizant partners, puppets and puppeteers. This balance is not aimed to support the voice of a single designer, user, or performer. Instead, it enables a critical dialogue. The discussion closes by emphasizing the role of this critical encounter in maintaining an adversarial yet productive stance.

The *Subway* project, a distributed dance performance, serves as the closing case for a combination of willed expression, material agency, and media participation.

Two Types of Oaks

Joseph Beuys remains a touchstone for twentieth-century art, an artist hard to miss. He was politically outspoken and was—among many other forms of engagement—one of the cofounders of Die Grünen, Germany's political party emphasizing ecological improvements at the core of its agenda. Eventually, he rose to become their candidate for the European Parliament election. Art and politics often merge in Beuys's practice. At times, the practice of political dialogue itself is the art.

Influenced by the educator Rudolph Steiner, Beuys developed the idea of a social sculpture (*soziale Plastik*) that framed both art and social engagement. One famous realization of a social sculpture started in 1982 for the *documenta 7* show in Kassel. *7000 Oaks (7000 Eichen)* remains a pivotal art piece within the canon of art history today. This long-running performance has been described as an artwork through "tree planting action," as an environmentally conscious urban reforestation project, and as a political and administrative work that demanded dialogue with city administration as well as the citizens of Kassel. Here, we do not discuss its positioning within the art canon but focus on the piece as an artistic happening in action, as an unfolding event involving many active parts. It "performs" on many levels while interconnecting them through the design of the overall piece.

Every five years, Kassel plays host for a hundred consecutive days to one of the most important contemporary art shows in Germany, *documenta*. Technically, the event grew out of a garden show in Kassel, a mainly industrial city that experienced massive bombing during World War II. It started in the mid-1950s as a developing platform to reconnect art in Germany to the modern developments. After the censorship that had limited art practices during Nazi rule, German art needed to develop its international connections and dialogues anew. The aftermath of silencing modern art as "Entartete Kunst" and the suppression of any nonconforming work had led to the isolation of artists in a nationalistic system and crippled artistic developments. The new republic had to overcome these confinements to invite new and experimental art back.

Beuys was no stranger to *documenta*. He had already contributed in 1977, when he participated in the first live video broadcast to celebrate the opening of *documenta 6*. But five years later, for *documenta 7*, Beuys rejected the idea of a fixed piece; instead, he was encouraged by the organizer, Rudi

Fuchs, to target a larger-scale public installation. Beuys focused on archetypical representations of nature, indigenous tree saplings, as a main component of the work. Another component of *7000 Oaks* were seven thousand basalt stones mined from Kassel's surroundings and the nearby Westerwald region. The stones were arranged on Kassel's Friedrichsplatz to form a giant triangle. The public was asked to adopt stones and trees for the price of 500 deutsche mark. Whenever a stone/tree was adopted, the tree would be planted in Kassel, and the basalt stone would be positioned in front of it. Beuys himself planted the first tree. His son Wenzel planted the last one in 1987. Many others adopted trees in between, and even after the official last sapling was planted, individual planting actions saw trees appear in new locations. The wide audience participation reflected one of Beuys's most famous quotes: "Everybody is an artist." The core activity of the piece centered on the purchase of stones and the planting of trees by and for the public. Progress of this action was reflected in the number of newly planted trees spreading through Kassel with the corresponding basalt stones positioned in front of them. It created thousands of smaller local installations. It was also visible in the gradually shrinking stone pile on the central Friedrichsplatz. But those were merely external indicators of the emerging dialogue that would surround the event.

The project posed many challenges from its planning to the financing to the realization to its maintenance. Given the nature of *7000 Oaks* as a social sculpture, all these challenges became part of the overall piece. Beuys was most aware of this and turned many of these components into artistic performances themselves. In that way, even the financing of the project was staged as an event. Financial support was provided by the Dia Art Foundation in New York, but to attract additional funding, Beuys staged happenings to attract more funds. These included auctions of work donated by other artists and himself and the creation of new pieces. For one of them, Beuys purchased a replica gold-cast crown of Ivan the Terrible. The replica had been constructed by the goldsmith René Kern in 1961 and had eventually become a showpiece in an upscale bar in Düsseldorf, where it was occasionally used as a drinking chalice for high-paying customers. Beuys announced his intent to destroy this symbol of the past empire in a melting ceremony and transform it into a newly cast piece. In the build-up to the crown's destruction, Beuys stirred up as much media attention as possible through public exhibitions and personal appearances, explaining that he would melt this symbol of past emperors and elitist decadence and cast a new *Friedenshase* (Peace bunny) statue. The *Friedenshase* was sold to Joseph Froehlich, an amateur art enthusiast, who later became a major collector. In return, Froehlich cofinanced the *7000 Oaks* project. In another financing effort, Beuys posed for posters and recorded a TV spot in an advertising campaign for Japanese whiskey. Media of any available kind, from local window exhibits of the gold crown to international TV spots, were part of the project even before the first physical tree was ever planted. A dialogue had started; newspapers and magazines ran stories, and public responses were pouring in. For example, the destruction of the crown replica turned into a spectacle itself.

As a performative work in a public space, *7000 Oaks* required complex negotiations with city administration and planners before the massive planting activity could start. Even the project's slogan, "Stadtverwaldung statt Stadtverwaltung" (City forestation instead of city administration), used a wordplay to bring attention to these administrative tasks. Project administration was handled by a dedicated group overseeing the stone purchases and tree installations. The whole process stretched over five years—outliving Beuys himself, who died in 1986. But even the official end of the planting action in 1987 did not close the piece. The maintenance and development of the piece and the care for the trees have been the task of multiple foundations ever since and have been central to public debates. The trees and stones developed as active partners in a larger discourse.

On July 19, 1984, a motorcyclist crashed into a car, which pushed him into one of the basalt stones now spread all over Kassel. The impact killed him instantly. In response, local press criticized the basalt stones as "death traps" (*Todesfallen*) and requested their removal. That did not happen; instead, the city removed the commemorative cross installed by the family but left the basalt stone in place. In other cases, vandalism led to the destruction of trees or graffiti defaced the basalt stones on Friedrichsplatz and other places. In these ways, the piece entered a complex and multifaceted public dialogue. It fulfilled its purpose as social sculpture, and it continues to do so today. All the while, the trees remain living components of the artwork as they grow, pollinate, get sick, or die off. The project remains problematic, alive, and unresolved.

7000 Oaks inspired many other social sculptures, spin-off plantings, and debates. Among them is a virtual reenactment. To honor its twenty-fifth anniversary in 2007, the work was reenacted by Eva & Franco Mattes in the virtual world of *Second Life*. *Second Life* is an online multiuser environment launched in 2003 by the San Francisco–based company Linden Labs. It allows participants to create their own avatars, log on to various public and private virtual spaces, explore a three-dimensional virtual world, interact with other users in this virtual world, purchase elements in it, and create and sell their own virtual objects or real estate.

Eva & Franco Mattes, collectively known as 0100101110101101.org, staged their virtual version of *7000 Oaks* on the *Cosmos Island* in *Second Life*. To the day, twenty-five years after the first tree had been planted by Beuys, they installed their first virtual tree and virtual stone in an public online event. The performance was part of a series of virtual reenactments of seminal performance artwork in *Second Life*. Other reenactments included Chris Burden's *Shoot* and Marina Abramović and Ulay's *Imponderabilia*. Part of their motivation for this reenacting was a problem with performance art as such, as Franco Mattes noted: "Eva and me, we hate performance art, we never quite got the point. So, we wanted to understand what made it so un-interesting to us, and reenacting these performances was the best way to figure it out" (Franco Mattes qtd. in Caronia, Janša, and Quaranta 2014, 99).

As long as one owns a computer and an internet connection, *Second Life* is a readily available and configurable multiuser online world, an experimental petri dish to "figure it out." In such an online world, "everything is mediated, nothing is spontaneous. More or less the opposite of what performance art is supposed to be" (Mattes & Mattes 2007). The total mediation of the online environment is at once the opposite of performance art's in-the-moment creativity, and at the same time, it offers a laboratory. Here, "there's no distinction between reality and fiction, facts and fantasy, authentic and simulated. Nothing is real, everything is possible" (Mattes & Mattes 2007). In that way, the virtual reenactment is more a critical encounter than a historic restaging. It changes its nature through the new format. In its differences and parallels, it allows for a critical reflection, and a range of differences stand out.

In contrast to the complicated origin story of the original piece and its struggle for financial support, the virtual performance was commissioned by a single source: the Spanish Centro de Arte Juan Ismael. As seen in figure 3.1, the initial virtual stones piled up in a virtual online space with only a basic model of the Fridericianum, the museum building that dominates the original

Figure 3.1
Reenactment of Joseph Beuys's *7000 Oaks* (2007), Eva & Franco Mattes. Used with permission by Eva & Franco Mattes.

Friedrichsplatz in Kassel. Beyond it, there is no cityscape that needed reforestation in the first place. *Second Life* splits its virtual real estate into "islands" and the Mattes's island, named *Cosmos Island*, was customized specifically for the performance. The reality of the rendered image did not recreate the actual architecture, streets, houses, or citizens but worked on its own terms. This space was accessible globally, visitors could teleport into it, or fly over it. Given the nature of *Second Life*, participants visited by steering their avatars to *Cosmos Island*, and there they interacted through prepared digital objects. Tree objects were free for the taking to visiting avatars. Each one was coded as its own object that could be dropped anywhere in the *Second Life* universe depending only on the ownership status of the chosen virtual plot. The design provided a unique digital representation with specific material conditions. The "social and the material are entwined" so that the "materiality of the digital shapes the cultural experience we can have of it and the purpose to which we put it" (Dourish 2017, 72–73). But these conditions differed clearly from the ones Beuys had faced before. There was no city administration to deal with; neither was there a city. The growth of a virtual tree did not require much financial commitment or planning; nor did it depend on the right soil, sunlight, or surrounding architecture and traffic conditions. No organizations or foundations were needed to maintain the planted trees as their own objects distinct from the *Second Life* world. No debates ensued over their presence in the neighborhood. It was as easy to retract virtual trees as it was to drop them at a location in *Second Life* or remove them once again from there. Virtual trees have no roots that could tie them to any particular location. Once the participants had left the Mattes's *Cosmos Island* with their virtual tree giveaway, those tree objects were not tracked by Eva & Franco Mattes but left to disperse into the sprawl of the virtual world. Still, the objects themselves contained ownership information that remained encoded as metadata into each object populating the virtual world of *Second Life*. Clicking on the tree object would inform a visitor about this object's personal ownership information as well as the object's origin. It seems that not all virtual trees/basalt stones had been collected even though the virtual invitation to participants stated that "the project will end once 7000 virtual stones and trees will be placed all around Second Life" (event invitation cited from Ninsve 2007).

As inviting as a free online world might sound, it inherently builds on many limitations. The software itself poses technological limitations, such as the server performance that supports limited numbers of visiting users to any single island, as well as limitations to the graphic display, textures, and shading of the world. Another restrictive condition of the *Second Life* platform is that the virtual world establishes its own legal limitations in the form of the End User License Agreement (EULA). These are the terms of service that all users of the *Second Life* software have to accept to run the software. Not unlike Beuys's discussions with city administration, interest groups, and lawyers, the reenactment had to operate under the conditions of certain laws, the EULA of *Second Life*. In *Second Life*, however, any such legal discourse is decidedly more one-sided. To continue to participate in the virtual world, users depend on the permissions and features provided by its operator. The owner of the *Second Life* system can—and repeatedly does—change the EULA. Such changes are foundational as the terms of its use regulate how one can use the technology itself. They affect conditions for commercial use and in-game ownership of objects, for example.

Ultimately, these limitations affected the life span of the work and its activity. Even though the digital information was stored on central servers and shared globally, the virtual location of the reenacted piece, *Cosmos Island*, is no longer online. The virtual Fridericianum is gone, and so is the entire origin space. It is hard to tell whether any of the virtual trees still exist in some corner of the *Second Life* universe or whether the reenactment left any other traces.

The virtual reenactment also caused a lot less controversy than the original Beuys piece. The artists mention disgruntled email responses from "real world" performance art enthusiasts (Mattes & Mattes 2007), yet the most substantial critique might have been one that tackled the underlying technology. *Second Life*'s inherent focus on a detached virtual alternative world stands in contrast to the original concept of a piece rooted in environmentalism. *Second Life*'s servers, the connecting network, and the end-user computers all depend on electricity and functional information infrastructure. In his critique, Patrick Lichty (2009) contrasts the original concept of reforestation in Beuys's piece to cursory calculations of actual energy costs of any *Second Life* avatar and island. From this point of view, the actual environmental impact of the digital infrastructure counters the reforestation aims of Beuys's original piece as it depends on energy consumption without providing any relief to the environment. Instead of promoting forestation, the energy needed by the networks that operate the foundations of the virtual world required energy consumption without providing any counterbalance.

Finally, the lack of financial continuity can erase the project entirely, especially the disappearance of the virtual origin space, which frames the reenactment as a temporary comment in contrast to the ongoing scale of Beuys's original event. In the original design of *7000 Oaks*, Beuys contradicts the relative stability and consistency of the basalt monolith with the changing biological mass of the adjacent growing tree. Both exist in relation and as a pair that engages with the public to support the original social sculpture. In its virtual reenactment, this conversation is muted by the digital nature of *Second Life*. Spaces and objects disappear and cannot maintain longer discourses in the same way.

Even though the virtual *7000 Oaks* reenactment uses a social media platform and allowed visitors from all over the world to encounter the piece, even though it had the powers of internet communication at its disposal for distribution and discussion, it failed as a social sculpture of permanence.

The original island is gone; the virtual trees (if there are any left) remain unaccounted for. The reenactment remains a powerful artistic experiment by Eva & Franco Mattes, but the project has not established itself as a dynamic part of the social fabric of this virtual world. For an exploratory conceptual piece, this might never have been the main goal, but it highlights the new role of the computer as media and its shortcomings. The Matteses' reenactment includes moments of emergence but ultimately diffuses into the conditions of the platform. That platform, however, provides inspiration for further development and reflection.

"While working on the reenactments, we realized that some of the best moments were software errors. For example, bodies were merging into one another. It shouldn't be like that and sooner or later, they'll fix it. So we got interested in these errors and thought that a 'videogame-native' performance should work on this and include software bugs. We're not playing the computer like you play a guitar; we're playing with the machine, we get influenced by it. The computer is not an instrument; it's more like a partner" (Eva & Franco Mattes qtd. in Shindler 2010). In that way, the Matteses' reenactment includes the computer as a performative partner with its own limitations as a social technology but one that still provides the power to contribute to creative action. The reenactment might have failed in the long run as a virtual social sculpture in *Second Life*, but it worked as an experiment for the Matteses' technological encounter.

Entering Performance

Performance is a rich, multilayered, and problematic lens for media studies. Scholars readily adopt different interpretations when using performance-driven approaches and deduce varying conclusions. Performance studies itself emerged as an umbrella discipline to engulf and expand drama studies. It builds on neighboring disciplines, such as communication studies and anthropology but reaches into many other areas. This reach across domains requires some clarification about how it may be used here.

The term *performance studies* emerged in the 1960s as an expansion of traditional studies of drama and theater. Richard Schechner, one of the driving forces in this expansion, initially connected performance to personal and social spheres that range from sex to sport to ritual. He considers it to be an "inclusive term" (Schechner 2003, 9) that defies limitations to a single unified field. Consequently, the critical study of performance art, performance studies, is dealing with a "self-contradictory" (Schechner 2002, 22) and open matter. Schechner (2002, 30) distinguishes between an event that "is" performance—framed and announced as such—and one that might be seen "as" performance, wherein performance offers a lens, a "means to investigate what the object does, how it interacts with other objects or beings, and how it relates to other objects or beings."

We can see further expansion of such frames in the work of scholars like Jon McKenzie, who uses the term in an even wider sense. McKenzie differentiates between three main forms of performances. First, *organizational performance* describes the operation and labor of employees who produce for a company or entity. Second, *cultural performance* describes "the living, embodied expression of cultural traditions and transformations" (McKenzie 2001, 8), which is framed by theater stages, fenced rock concerts, or sport arenas. The third, *technological performance*, relates to the operation of machines. He traces this technological performance in the operation of objects from assembly lines to plant seeds but particularly in computers and their emergence during the Cold War period. Objects are seen as performing but only through the eyes of the human. Missile-guiding systems perform but only as actors in the Cold War raging between the superpowers.

Beuys's *7000 Oaks* performs on all three levels: from the dealings with the local city administration, to the specifically framed events of the plantings, to the operation of the trees in the current cityscape as they shape the unfolding life of Kassel. Simondon's steam engine performs on all three levels: from the Industrial Revolution caused by their integration into factories, to the spectacle of the steam engine across many media (starring in the earliest film recordings, for example), to the concretization of their inherent technical operation. And digital media perform on all three levels. They form the basis for new forms of digital labor that build on the contributions of often widely distributed participants, as seen in the "crowdworking" Mechanical Turk by Amazon. They deliver networked performances that see musicians live coding pieces as they unfold, both on location or distributed over the internet. They support established fields framed as artistic practice, such as digital art, which include exhibitions, programs, publications, and such events as the *7000 Oaks* reenactment.

McKenzie's reach in performative practice and Schechner's discipline-blurring perspective open up performance as a field, and it is not surprising that we find varying traces of a "performative turn" throughout digital media and its debates. Brenda Laurel (1991) uses performative models to shape the design of digital experiences. Other scholars turn more to the technological part in the performance: Alexander Galloway (2006, 2) emphasizes the role of the code as the game is being "enacted," for example. Performance is a shared quality even between otherwise conflicting scholarly approaches. For example, Espen Aarseth (1997) speaks of a "performative character" in video games as textual machines that produce text in a collaboration of machine, sign, and interactor. Even though they fundamentally disagree on their view on videogames, Aarseth's turn to performance relates to Janet Murray's (1997) notion of "cyberdrama." Both emphasize the role of the player in the construction of the game as emerging text, while they differ in their stance to neighboring media, such as literature, film, or television. The interactor turns into the "inter-actor," a term adapted from Augusto Boal (2000) and his *Theater of the Oppressed*. New media, such as video games or online worlds, call for novel forms of this inter-actor's integration.

This inherently concerns the way we design the participator's interaction with the computer at hand. Not surprisingly, the neighboring domain of human computer interaction picked up performance to inform the theory and practice of interaction design (Benford and Giannachi 2011; Chatzichristodoulou, Jefferies, and Zerihan 2009; Jacucci 2004). As the digital spread further into increasingly hybrid conditions, these practices grew ever more complex when dealing with ubiquitous digital media, such as cell phones or the internet of things. Performance was already a flexible domain when the interaction domain adopted it, but the ongoing spreading of digital media through different outlets and technologies added a further level of complexity.

At the same time, and in conversation with such technological changes, digital media continue to shape performance and affect the debates in the field of performance studies and art (Causey 2006; Dixon 2007; Salter 2010) as they are being remolded by the digital. New techniques allow for novel forms of performance practice and blur established categories. Among them is the question of what constitutes a "live" performance, or how to draw lines between different media in a digital age. This did not start with the digital. Principles of liveness and intermediality are not by definition digital (Auslander 2008; Chapple and Kattenbelt 2006). But the spread of digital media accelerated and diversified this development as it provided versatile and accessible digital toolboxes. They include computer graphic imagery, digital sound manipulation, networked performances, online worlds, and robotic actors, among many other new opportunities. Performance has informed the development of these tools on the side of media technologies, for example, in debates on video game textuality or interaction design. But the results have also shaped emergent performative practices through new technologies on and off the stage. This proliferation of performance into different fields and the blending of new technologies into performance practices highlight the value of the dialogue between performance and digital media. At the same time, the fluid encounter and the sheer range of dependencies require a clear positioning on how an argument for vital media will use performance and how it will discuss the many contact points between these domains.

Expression and Action

For the purposes of this argument, performance is defined as centered on the moment of individual creation of action—namely, as *the action to produce expression with an intent*. This builds on Erving Goffman's (1956, 8) formulation: "[A] 'performance' may be defined as all the activity of a given participant on a given occasion which serves to influence in any way any of the other participants." Goffman emphasizes action as much as the role of an audience and an effect that "influences." His definition builds on face-to-face interactions between participants "in one another's continuous presence" (8). It is based on a situation, "an occasion," as Goffman would call it. In contrast, the working definition used here—*action to produce expression with an intent*—focuses on the performer as producer first. It still centers on the moment of production but does not yet look at the effects of this produced action.

As with every other action, performance is a cocreative process birthed by the formative activities of many partners and emerging through many operating agencies in an intra-active emerging network. It is the "expression" component that distinguishes it from most other activities. To clarify where this expression comes from and how it is included in the performance action, we need to return to the cognitive operation.

Expression refers to the cognitive construction processes that the human contributes and that are meant to carry significance beyond mere function.

As humans evolve their cognitive abilities in cooperation with their surroundings, these abilities probe and explore. They help us to critically reflect and creatively find new ways to articulate ourselves and shared ideas. They experiment and provide their own activities to the networks they are part of. Following Welsch, the cognitive evolution stems out of a sensing, processing, and activating historical development. But a willed expression differs from other cognitive activities. It does not refer to basic sensory activations, such as a gasp when one bumps into a table corner on the way to the kitchen or the instinctive reaction of raising your hand when a ball is kicked toward your face. Instead, the activities we are looking for are willed. That means that they not only produce a physical action outcome, but they also aim to further that cognitive construction process. These are not mere reactions of the sensory system but active interpretations, explorations, and experiments. When we strike a pose or break into a dance step, we do not simply navigate, but we do so with a willed quality to it. It is one thing to drop into your chair with the first cup of coffee and the mind barely conscious. It is another thing to find yourself at a table in a fancy restaurant picking from the menu selection. Goffman would rightly highlight the different frames that such situations generate. In our case, the difference focuses less on the social context and more on the differing cognitive awareness of the human actor.

This performance production is initially centered on the maker, but even though the following sections will keep this focus firmly in place, the importance for communication and social construction should be noted early on. The impact of a gesture can be immense and can construct new framings itself. But the larger social constructive work of performance remains an undercurrent until questions of such cultural production will resurface in the last chapter of the book. What is key for the notion of expression at this moment is how we differ in the production of our expressive actions, our gestures, movements, words, or other forms of activity that include a willed cognitive component.

The second point that needs clarification is the differentiation between activity and expression. Activity is not limited to cognitive beings—expression is. Action is open to anyone and anything. An earthquake, a combustion motor, or a robotic arm can act. In contrast, expression as a productive force depends on interpretation and sentience. It depends on a will to act in a particular way, based on a particular experience, and as a result of a reflection concerning the situation at hand. Expression goes beyond an immediate function and adds a kind of action that not only fulfills the worldly necessities but also furthers the cognitive relation to them. Developing and expressing such an expressive intent is necessary for us to become individuals. We produce presentations within social conditions (Goffman 1956), and we produce these social conditions through our participation. The "stage drama" informs the "social drama" and vice versa (Turner 1985). We perform and construct power relationships, political hierarchies, and categories, such as gender (Butler 1990). Performance shapes cognition, and performance is tied to cognitive beings as they engage with and build the world around them. In our case, then, cognition calls for expression, and only cognitive beings need expression beyond biological functionalities.

Only cognitive beings have developed such performative expression practices because only they need them to support identities. It is a curse as much as a gift, but most importantly it is a necessity. "Identities are a key means through which people care about and care for what is going on around them. They are important bases from which people create new activities, new worlds, and new ways of being" (Holland et al. 1998, 5). Expression and performance are needed for the construction of identity in relation to the social and nonsocial surroundings.

Noncognitive actors do not depend on these expressive means in their development but focus instead on their own operational dependencies that are equally important but qualitatively different. Abilities and needs remain balanced in the evolutionary cycles that we form in our collaborations—or at least they should. Within this universe of emergence sits the sentient being, who requires and performs its own activities that distinguish it from the surrounding not only through means of physical but also cognitively expressed difference.

Making personal identity is always also part of a social construction. Others might be influenced by the activity, and situations change. Wearing a wedding ring, or a crown, or handcuffs sets some clear social signals that can affect how others relate. The focus on such social construction is important and much debated. Vital media aim somewhat lower as they hope to rebalance humans' relationship to the operational material surroundings. That is why we turn our attention to the initial action performance and its coconstruction in relation to the surrounding materials. Any such situation has a social component to it, but for the time being, we will focus more on the encounter with the noncognizant material than on the social impact of any such performed activity.

This logic splits the traditional view of performance as communication into a two-step event: First, an event is constructed in collaboration with surrounding world components, human, nonhuman, analog, digital, or any other kind. This production includes the element of willed expression originating from the human coconstructor as well as the activity of other operators in the network of activity that allow each other's coming into being. Second, this event will have certain effects in this world beyond the immediate moment of its construction. Whoever constructs the expression might intend to speak to an audience, and this impact can be important. However, it is not inherently necessary to succeed in the original production process. An expression might very well be ignored or misunderstood and remain a performative act. Any such action unfolds in a social sphere, historic, or cultural condition, but in the turn away from a human-centered media model proposed here, we turn from this sphere to the smaller, more immediate material production moment.

That means that a performative action might very well fall flat as an outward-directed communicative act but remain a successful performative action for the cognitive and noncognizant constructors based on our working definition. The actors collaboratively manage to complete the production of the expressive action, and this concludes the most basic condition. At this moment, the question of its success on the receiving social end of the action is secondary. This detaches our concept of performative action from another dominating view: that of performance as effective means of communication between humans. An expression might very well construct a successful communication process, as might be the case with a successful theater performance, a lively stand-up show, an effective classroom session, or a video game play session. But the initial production of performative action is not dependent on its success with other humans. It is the collaboration of the human with surrounding materials to produce that willed expression. Our working definition of performance—the action to produce expression with an intent—centers on that moment of coproducing the originating action.

Limited as this production-based view is, it separates performance action from nonperformance action: the former includes the intent to express; the latter lacks this intent and the targeted expression. The first implies the opening of a possible dialogue regarding that expression—whether it manifests is a different story. The second does not even aspire to do so—but it still might be interpreted as such. Neither can fully guarantee any successful dialogue or effective influence.

Digesting one's morning coffee is an activity, so is driving a car or preparing said morning coffee. But unless there is a willed expression, these are not performances. They can be performative: stunt drivers use their skills and optimized machines to construct a performance; a skilled barista can turn the preparation of a cup of coffee into an art form. But these activities differ because a stunt driver or barista uses them as expression through their specialized abilities. They define themselves and grow through their practices. If neither of them is needed for us in terms of self-definition, then we readily allow them to be outsourced to automatic coffee machines or self-driving cars. Thus, key concepts in our take on performative action are action unfolding in a combination of willed expression and noncognizant activity. The result has to be a form of cognitive change, as well as a material one, in the producing actors.

Functions that do not inherently depend on willed expression or lack changes in state can often be handed off to nonsentient actors. In return, this can lead to a lack of relation as these automated actors can mute expressions and reflections that could otherwise seep into parts of the activity. The "material drama" loses its ability to inform the "social" one. The result can turn into a commodity, not because it involves material actors but because it lacks expression and individual growth and connection. Borgmann (1984) calls this the "device paradigm." As technology makes life easier and reduces human involvement, it turns experiential "things," such as a wood stove that requires attention and skill to operate, into "devices," such as a central heating system that provides warmth but "makes no demand on our skill, strength, or attention" (Borgmann 1984, 42). Engagement is reduced; cognitive investment is lowered "as a central heating plant procures mere warmth and disburdens us of all other elements" (1984, 42). Here, action displaces and flattens out expression. "The device provides social disburdenment, i.e. anonymity" (1984, 44). The operation of a thermostat was one of the early examples for systems of control in the development of cybernetics (Wiener [1961] 2019). Handing off the mechanisms of control identified by Wiener causes Borgmann to lament the disburdenment.

For our media-balancing act, the question is not whether action is more important than expression, whether the agency of noncognitive actors is more relevant than that of cognitive ones. It is the realization that the necessary role of the human in the mediascape is that of an engaged and expressive one and that any such expression is dependent on material agency. As a consequence, the richness of the encounter is more important than simplifying access or optimized control mechanisms. Just as a social perspective of performance led us to break open hierarchies and social categories, a material perspective should invite us to break open categories of control and commodity. Vital media build on this kind of critical coproduction. They are networks of actions that emphasize the interrelations of forces to counter disconnects and anonymity. In vital media, there is no productive expressive action without material agency. At the same time, the action cannot replace our own investment of willed expression. This is the performance-based functionality of vital media.

Scream with Machines

Kelly Dobson's (2008) father and grandfather both worked at a junkyard, she herself used to work there, and later she noted that "my family has a lot of expressions that are like grunts and groans, that are meaningful but that they have learned from the machines." During her studies, Dobson found herself visiting large building sites and harmonizing, vocally resonating, with the sounds of machines. Eventually, she replicated the effect in a lab. She even started to invite others to join and encounter machines in such a personal way in special lab spaces, where she created encounters she termed "machine therapy." One project that emerged from these experiences was *Blendie* (Dobson 2004). *Blendie* appears to be a stylish 1950s Osterizer blender. However, instead of operating this blender with buttons, one has to imitate its machinic sounds to activate it. Dobson hacked the original blender, included a voice control feature, and trained the system with human voices imitating blender sounds. When activated, the sounds of human "grunts and groans" mix with the blenders own machinic sounds when it starts its motor and adjusts the rotation speed to whatever frequency the human voice suggests. *Blendie* is part of a series of explorations that saw Dobson experimenting with vacuum cleaners, drills, or sewing machine sounds as she tested designs that would support empathic relationships with machines.

By bringing the sound of the blender to the foreground, we realize an operational aspect of the machine that might otherwise find less attention, that might be shut out as a nuisance, much like we try to ignore the sounds of building machines in construction sites. Dobson's design of *Blendie* counters such avoidance, and it does so through performative activation. The human screams at the blender, imitating its sounds, recognizing the blender's originally disturbing acoustics, enhancing and mimicking them, as the blender responds with its own operations. The design emphasizes the blender's material agency. At times, the machine even activates without human input to surprise its human coperformer. One clear sign of the emerging relationship that this design supports is that the human performance goes beyond the acoustic scream.

In one video (see figure 3.2), Dobson demonstrates the functionality of *Blendie* as she pops up from under the table to approach the machine, eyeing the blender, clearly addressing it in gesture and pose even before she vocalizes her scream. Eventually, she crawls on top of the table while grunting the

Figure 3.2
Blendie in action as part of Dobson's *Toward Machine Therapy* (Dobson 2004). Reprinted by permission from Springer.

machine into activity. The technology embedded in the machine still only picks up the sound input and does not recognize any movement or gesture beyond that. The fact that Dobson saw fit to include dramatic movements and facial expressions shows that the empathic relationship goes beyond the vocalization, and the human performer produces expressive action beyond the necessary sounds. "The action may also bring about personal revelations in the participant, because in sounding with the blender one is likely to perform gesture and sound expressions not previously accessed which may open up unfound emotions or thoughts or feelings" (Dobson 2004). The success of *Blendie* is based on its material-based design, which stimulates human performance in relation to the object. The blender's original noncognizant function still operates as expected. It still mixes drinks. Its 1950s appearance puts it squarely into predigital aesthetics that hide its machine-learning and embedded-computing components. Instead, it highlights traditional functionality. At the same time, it calls the human to adjust and to stretch their expressive range to relate to the machine.

As a design project, *Blendie* pushes the machinic actions of the technical object to the foreground. This does not detach the machine from the human who engages with it. Its human-machine interaction might be guttural, but, as Dobson noted, it supports an expansion of one's expressive range, one's cognitive engagement. It invites and supports further actions, as the dramatic encounter through movement in Dobson's own video illustrates. By design, *Blendie* counters a commodity-based disconnect. It, furthermore, does not define the success of a human performance as a social event for others but as the personal activation of the machine. As a public project, we encounter it only as presented to us in talks, papers, art shows, or other installations. In these settings, *Blendie* becomes part of a framed cultural event. But at its core, it remains an extension of Dobson's encounters with machines on junkyards and building sites. Any such encounter unfolds within a social context, but equally clearly *Blendie* centers on the encounter with the object first. It operates in the logic of machine therapy. It does not focus on the social construction—the emerging role of the construction worker or the cook in their communities. Instead, it highlights the relationship to the machines on a construction site, the blender on the kitchen table. Most of the time, these machines are operated without any vital media realizations. The way most blenders are used is most likely not a performance but an almost automatic step in food preparation. But

Blendie can, by design, only be operated in a performative way. In that way, it reconnects to the material object and cognitively activates the human participant. It complicates the "normal" mere functionality to stimulate the human participant. It expands the agencies of the machine and the cognitive engagement of the human in combination.

The connectedness of humans with nonhuman elements remains fundamental as performance emerges from the evolutionary condition of coexistence and the design shifts its focus on this productive encounter. *Blendie's* design brings forth this context—other digital designs suppress it. The detachment of the virtual trees from the reenactment of the original quality of *7000 Oaks* is one example of such a suppression. Compared to the original Beuys piece, the reenactment misses the encounter of a human's willed expression with the object's own activity. The virtual instance centers on a different encounter limited by the conditions and operations of the *Second Life* platform. The digital tree is not devoid of materiality, but both the virtual tree's activity (its embedded code, possible animations) and the performance of the *Second Life* platform (its networking, rendering, interaction design) are commodified instead of brought forward. In contrast, Beuys's idea of planting biological trees as an act to form a social sculpture presents a process toward vital media that still unfolds in the city of Kassel. Dobson's interaction design emerged from a different urban problem: building sites. *Blendie* focuses on the personal material encounter itself. It is less a social sculpture than a mediated object performance. Just as well, it pushes the performance with the active object at hand to the forefront to counter any commodification.

Building on the foundations of collaboration and cocreation, this design approach opens opportunities for activities between humans and nonhumans. Performance allows them to codevelop opportunities between cognitive actors and their noncognizant surroundings. For humans, as they perform, this means that performance action can serve self-differentiation and self-construction. Through this activity, all actors, trees, blender, and human rise out of Borgmann's "anonymity." This expression is not a seclusion from other objects but provides means to construct a meaningful connectedness and coexistence for cognitive beings in relation to other cognitive as well as noncognitive ones. This kind of performative action still is a productive act—an evolutionary step and probing of the inner personal possibilities as much as the outer physical ones. Vital media design calls for such enriching encounters to establish the expressive collaboration.

The current examples emphasize physical things and their voices. But objects and our encounters with them are often mediated through television, film, or digital channels. How do these forms of mediations affect the performative production process of vital media?

Mediation in the Digital Age

The differences between virtual and biological trees might serve to highlight the different yet interdependent forms of agency, but it does not tell us where these digital media objects stand themselves nor what their impact is for performative functionality. This section turns to such digital objects and their particular role in the mediation processes.

Goffman's (1956, 8) definition of performance centered on the idea of "the activity of a given participant on a given occasion" aimed to influence other participants. It regards the condition, or "occasion," of a performance. Goffman sees any such performance as only possible in the state of the "continuous" and "immediate physical presence" of the participants as they influence each other (1956, 8). This kind of immediate encounter is threatened by mediation. Mediation has evolved into a ubiquitous activity in the digital age. As we perform everyday life and our interactions with other humans, we have also become "entangled" (Salter 2010) with algorithms, machines, and objects. Chris Salter (2010) looks into the ways that technology has changed practices and meanings in performance and how this has created new opportunities for artistic work. Salter's (2010, xxxiii) idea of entanglement "suggests that human and technical beings and processes are so intimately bound up in a conglomeration of relations that it makes it difficult, if not impossible to tease out separate essences for each." As we perform together with technology, we blur the lines between the actors involved. Simondon's "milieu" has already served as one approach to trace these entanglements. But Simondon was interested in the question of production. In contrast, we step with Salter into the challenge of digital mediation and performance.

Digital media challenges Goffman's definition on multiple levels. His fixation on shared physical presence cannot apply anymore in a digital ecology of distributed actions and participants. Online performances like the Matteses' reenactment, video game play sessions, robotic and hybrid performers, or multimedia concerts question the value of physical presence. Digital media partners can stand in between humans and machines, involving, connecting, separating, activating, engaging them. Signals and networks become part of the performance and complicate the localization of a single event that might be broadcasted, recorded, or streamed.

Philip Auslander (2008, 24) has provocatively argued that the dominance of television in modern culture "has enabled it to displace and replace live performance in a wide variety of cultural contexts." Performance today, he argues, necessarily deals with mediation, either deploying it actively or remaining aware and shaped by its presence, aesthetics, and systems. He draws media into performative action as he reads live and mediatized performances as related. "I might argue that live performance has indeed been pried from its shell and that all performance modes, live or mediatized, are now equal: none is perceived as auratic or authentic; the live performance is just one more reproduction of a given text or one more reproducible text" (2008, 55). For him, the TV image production presents a performative act. It might have the makings of a disembodied mass media, but, Auslander argues, it provides an "experience of evanescence" (55). Its production is no less "live" than any human performance. In practice, these kinds of mediated events have taken over and become the reference in performance. This focus on mediation questions the human performance situation, and this questioning is not unlike the focus on material encounters that vital media suggest. But Auslander does not build on material agency in his argument. He is more driven by a correction of preeminent views of performance as an exclusively human activity in an age of media. Still, his inclusion of media technology in performance corresponds to the shift that is proposed here. Both share a turn away from a humancentric view, though for different reasons.

In Auslander's mediatized age, no performance can be unaffected, and this clashes with views of other performance study scholars, such as Peggy Phelan (1993), who sees the presence of the "living" human body as central to any performative act: "Performance's only life is in the present. Performance cannot be saved, recorded, documented, or otherwise participate in the circulation of representations *of* representations: once it does so, it becomes something other than performance. To the degree that performance attempts to enter the economy of reproduction it betrays and lessens the promise of its own ontology. Performance's being, like the ontology of subjectivity proposed here, becomes itself through disappearance" (1993, 146; emphasis in the original).

Vital media read performance in between the positions of Auslander and Phelan. First, "performance's being"—or better, it's becoming—is not anymore exclusively bound to human agency but a collaboration, merely involving humans. Second, this becoming has been recontextualized and reshaped through digital technology and countless media performance activities. Performance remains grounded in the contribution of the human through dependency on cognitive action, but performance is also intra-active in the way it incorporates material means. This includes mediating digital technologies. Such an evolutionary perspective does not have place for re-creation, only creation. In the network of actions that include performative ones, there is no re-production. Not even digital media objects can remake. We (humans, nonhumans, media) can only produce forward. Thus, the view on mediation here agrees with key elements of Phelan's position, such as the irreversible uniqueness of a performative act.

At the same time, material stuff is considered as active component and digital machines as equally important acting objects. We cannot reduce the performative situation to exclusively between humans sharing a theater stage. Instead, we should turn to performative materialism, an argument made on slightly different terms before by Barad (2003). Mediation is a form of such materialism. It is part of performance construction through its own material actions. That does not turn these performance processes into technologies of representation, but they remain actions of production. Participating objects and media cannot be re-creators because they never re-produce but can only act and evolve forward. Computers cannot undo a mediated moment; nor can we replay the same game moment twice.

Adding the digital to this construction process does not change the forward driving nature of production. But it changes the quality of these production conditions as it draws digital machines into the group of operators. What new materialism means to performance is a spreading of agency. What the inclusion of digital technology means is the addition of new forms of mediation into the vocabulary of those agencies. Here, "liveness" is indeed problematic because digital players have entered the stage. Yet, at the same time, performance remains tied to human contribution.

We should not take this expansion lightly. Among other effects, it complicates the notion of ownership, which diffuses in between agencies. Intended expression still depends on cognition, but the performance action itself is never single authored. It is distributed across human, material encounter and mediation practice. It is human dependent but not humancentric anymore.

This is where vital media's mediation practices and their approach to performance sit. The precondition of vital media is a realization of one's embeddedness in a hybrid world made up of tangible and intangible actors, cognizant and noncognizant ones, each one providing qualities to the surrounding and the unfolding actions. The materials, objects, environments, and atmospheres are engulfing, submerging, and collaborating with the human performer. They all contribute to the emergence of an intended expressive activity. As the cognitive components are contributed by the human participants, digital mediation adds its own qualities, such as networks, dislocated actions, even temporal distortions.

Acting with Digital Media

Focusing on a human-human relationship, Erika Fischer-Lichte (2008) introduced the concept of the "autopoietic feedback loop" to describe the coconstruction of performance between actor and audience. For Fischer-Lichte, performance depends on the bodily encounter between humans, some of them contributing as audience members, some as actors. The emerging action is always part of an ephemeral event shaped by that encounter, and it does not convey any pregiven meanings. Rather, the performance itself brings forth whatever connection might emerge. Central to her argument is the interoperation of audience and actor. "Whatever the actors do elicits a response from the spectators, which impacts on the entire performance. In this sense, performances are generated and determined by a self-referential and ever-changing feedback loop" (Fischer-Lichte 2008, 38). This feedback loop depends on ongoing contributions from both sides and constitutes the transformative powers at work in performance art. For Fischer-Lichte (2008, 12), transformations are not limited to a somehow mental or internal realization but "manifest in perceptible physical expressions—they flowed into and prompted concrete reactions" with clear consequences. Her examples include performances by Marina Abramović that included self-inflicted injuries to the artist and physically demanding relationships between Abramović as performer and "onlookers" that challenged the audience by design. Not only did this work change the designated performer, Abramović, as her cuttings left marks on her skin; it also altered the visitors to the performance, turning them into actors to inflict, tolerate, or prevent any such effects.

Famously, Abramović's *Rhythm 0* (1974) staged the artist as "the object" next to seventy-two other objects. In this object-verse, the audience members were encouraged to use any of those provided objects upon the performer's body. Equally famously, this object collection included a gun and a bullet, and the unfolding performance eventually led to a loaded gun in Abramović's hand, pointing at her head. Before that, numerous other physical actions had already been applied to her body by audience members. The physical danger her body was in, the objects at hand, and the behavior of the audience members were put in direct relation. The "audience" turns into active contributors acting upon a performer. There was no specific digital component in *Rhythm 0*, but its masterful interaction design still highlights emergent opportunities that can apply to digital performances. Events are risky, nonreversible, always evolving between all participants engaged.

This matches Barad's (2003) critique on representationalism in performance and her call to turn away from preimposed communication conditions between people. Barad (2003, 808) argues that we should instead look at actions and practices first: "A posthumanist account calls into question the givenness of the differential categories of 'human' and 'nonhuman,' examining the practices through which these differential boundaries are stabilized and destabilized." Vital media are not posthuman. The cognitive human actor remains a central and distinct part in the unfolding event and cannot be replaced by any noncognizant one, no matter how much agency this partner might be granted. But the intermingling of agencies and of destabilizing practices applies in digital vital media just as well.

Because of this intermingling, Fischer-Lichte's "autopoietic feedback loop" needs to be expanded in digital performances. It needs to include objects as much as media, math as much as guns. Actions between actors, such as clay, steel, acorns, or server farms hosting online worlds, remain on the same footing with any cognitive contribution. All are game in performance, and the digital has not detached whatever text is produced. It has, however, contributed its own specific qualities to the ongoing shared construction process. Unlike Abramović, we cannot claim to be objects without agency in such vital media production. Humans remain conscious contributors. Their willed expression remains necessary for the shared production. But they readily embrace the many other forces that help the events' unfolding. What would an AI have done during *Rhythm 0*? How would a robot have dealt with that gun?

Tackling such questions needs careful explorations of material agencies. If we project the dilemmas of cognizant beings onto digital or material ones, we do them injustice. Their actions are foundational, but they do not hinge on ethical standards and personal expression. Asking how a robot would produce a willed expression is like asking how a human would have dealt with the photosynthesis performed by the leaves of Beuys's *7000 Oaks*. Designs for vital media do not project human cognition onto the actions of machines but activate the different natures of both as well as their interoperation.

The focus on activities that are interdependent puts this call for interoperation closer to activity-centered design approaches. Largely driven by different readings of Lev Vygotsky's (1930) and Aleksei Leontiev's (1978) work, activity-centered design was already proposed in human computer interaction (Kaptelinin and Nardi 2006). It shares vital media's focus on the activity and the needs of individuals to construct themselves through these activities. The differentiation between object and subject remains central for activity theory, but it also recognizes their interdependence. "The very notion of interaction implies mutual effect produced by both sides on each other. However, the agency manifested by the subject of activities of a special character. It can be defined as *the ability and the need to act*" (Kaptelinin and Nardi 2006, 33; emphasis in the original). While it still focuses on the role of the human and social construction, it recognizes the importance of the object. For example, tools take a special place as they already embody much of the experience of humans in their design. Donald Norman (2005) has proposed it as a way to describe "slow, evolutionary folk design," which is a direction this book takes in the closing chapter. Where the call for vital media differs from both activity-centered design and the autopoietic feedback loop is in its inclusion of materials as equal contributors that help to bring all actors into existence. A good design manages to put the human right next to (and not above) these contributors.

Kacie Kinzer was not really interested in robots but on creating serendipity for human encounters, particularly in New York City, where she was studying at that time. Still, her project *Tweenbots* (2009) centered on a small robot that had only one mechanical operation: a motor to drive the little machine forward on its tiny wheels. The robot was made of cardboard, featuring a simple smiling face and a long stick with a little flag attached at its end (Kinzer 2010). One could find a message written on the flag. It called on the reader to assist the hapless robot in its mission to make it to a destination that Kinzer had also written on the flag. As the robot did not have its own wayfinding logic, it had no means to navigate by itself. Kinzer, then, released the robot in Washington Square Park and observed how numerous people walking through the park encountered the robot and assisted it to find its way from one end of Washington Square to the other. The design did not emphasize the sensing and decision-making processes of the robot but instead left these to the people who encountered the robot. That does not mean that the robot lacked its own contribution to this performance. It actively played its part. As expected, the robot constantly bumped into benches, stones, and obstacles. But it also included unexpected activity that unfolded as the robot encountered an environment that not only included humans. The day the robot was released in the park was very hot. It was so hot that one of the little rubber tires on the wheels of the robot stretched and came off. As a result, the robot broke with its designer's intended behavior and started to run in circles. It was trapped in that motion until two passing pedestrians figured out the problem and repaired the robot to send it back on its way (PopTechMedia 2010). *Rhythm 0* provided its participants with seventy-two objects and explored the emergence between them and the audience. *Tweenbot*'s park encounters were more open-ended by design and included the unpredictable environment of the public park with all its pitfalls and dangers for a hapless robot. As varied as these conditions might have been, somebody would always step forward and rescue the robot "every time the robot got caught under a park bench, ground futilely against a curb, or became trapped in a pothole" (MoMA 2011). Kinzer had selected the stage and included the robot as the main intervention within a complex network of conditions, and her design of the situation managed to successfully engage the park visitors in the overall effort. The entire event resembled a form of hidden performance or kinetic sculpture. Would the *Tweenbot* manage to make it from one corner of the park to another? What obstacles would it run into? How would the unaware human coperformers react? Ultimately, *Tweenbot* made it across the park. Not all robots are that lucky.

Some years after Kinzer's *Tweenbot* experiment, David Harris Smith and Frauke Zeller devised *hitchBOT*. Unlike *Tweenbot*, this robot was entirely incapable of moving by itself, yet it was set on even longer journeys. On its first outing, *hitchBOT*'s designers sent the robot hitchhiking across Canada. Left on the side of the road, it depended on passing drivers to stop and offer it a ride. Once it got picked up, it further relied on the humans to drop it off at the next location, where it would wait for the next driver. Like *Tweenbot*, *hitchBOT* relied on humans to succeed in its mission, and this dependency was central to the design, but unlike the tiny *Tweenbot*, the hitchhiking robot was a publicized spectacle. Its travels were covered by news teams, and the robot had its own Twitter account. Although *hitchBOT* was immobile, it featured more computational abilities than the *Tweenbot*. This included an onboard computer with voice recognition. With the help of Wikipedia, it could converse—on a basic level—with anybody picking it up. After successfully crossing Canada from coast to coast, a second instantiation of *hitchBOT* managed to make its way through Germany before moving on to the United States, where it eventually was vandalized and destroyed in Philadelphia, two weeks into its journey across the United States.

Both robots are often cited as examples of human-robot interaction and successful takes on social robotics. But here, they are understood as shared performances. The focus is not on how humanlike a robot can be designed but how the mechanical-digital bot can be used to emphasize a codependence in a performative event production. In this case, the destruction of *hitchBOT* is just another moment within its performance existence. It is a tragic one, but a successful one nevertheless.

Both robots ended up as ultimately demobilized exhibition pieces in museums. The first *Tweenbot* is at the collection of the Museum of Modern Art in New York, and the original *hitchBOT* has been purchased by the Canada Science and Technology Museum in Ottawa. The robots' agencies have ultimately been taken away as they have been turned into passive exhibition pieces to be looked at but not interacted with anymore.

If we turn away from a human-human perspective toward an intra-active one that also caters to digital components, then we have to deal with new dangers and opportunities. Both robot performance examples show design that supports this. The relative cuteness of these robots, their abstracted human faces and round shapes, might be engaging, but they are not the key components that count here. Instead, it is the emphasis on the robots' difference, their designated limitations that brought the robots' agency *and* the human one to the forefront. Vital media do not center on the human creating an emotional relationship to the robot or the robot mimicking the human as closely as possible. They do not call for noncognitive partners to be anthropomorphized. Instead, vital media center on the constructive encounter of the differences. This comes with risk, as *hitchBOT*'s untimely demise shows.

These robots invited the transformation of a passerby into a pathfinder or into a travel companion. Audiences turn into contributing actors. Murray (1997) made the case that such a "transformation" defines a key component of digital media. In her case, she derives it from the "shape-shifting" of the player of a video game. The player takes on a role and participates in an unfolding narrative provided by the designer in the game world. Murray realized that such a self-transformation happened to herself in a game arcade when playing the light gun shooter *Mad Dog McCree* (Roberts 1990). As she picked up the light gun from the cabinet, she felt immersed in the action and the dramatic position the game put her in. It awakened her culturally seeded identification with gunslingers such as Annie Oakley and Wyatt Earp (Murray 1997, 60). But the encounter of the gun in Abramović's *Rhythm 0*, the guidance provided by helpers to *Tweenbot*, and the destruction of *hitchBOT* are more pressing and less forgiving transformation results. Murray's idea of transformation emphasizes the player as central and their experience as the focal point for good game design. It is human centered with the material and digital affordances aligned to optimize the impact on the player alone. The light gun of the arcade game cabinet is only a means for stimulating the player's transformation. In contrast, vital media are relation based and codependent. They shape all participants—human, robot, or video game—and each item is necessarily altered in the process.

That is why vital media clash with another proposed feature of video games: vital media games cannot be replayed. They fall into what Schechner defined as "actual," and the idea of an "actual" transformation applies better to vital media than the idea of a human-centered experience design. What transforms is the situation with all the collaborators in it, not just one participant.

Schechner coined the term "actual" in his discussion of Allan Kaprow's happenings and through an anthropological view of performed rituals inspired by Victor Turner. In these cases, ritualistic performances are conducted that emphasize both, a mystery at the heart of the material action and "art as event" (Schechner 2003, 28), which is designed and produced to change the social fabric. For Schechner, these actuals include a mystic quality, which is experienced as a transcendental event. Yet the event itself remains concrete and organic in its striving for social and environmental wholeness. It can serve to rebalance the status quo in a society or to question it. It provides a performance-based struggle that supports the interdependent relationship of humans to nature and social realities. Not surprisingly, this includes initiation rituals as well as death rites, entry-to-world and exit-from-world moments. Describing actuals, Schechner draws on references to nature and evolutionary processes that are part of these passages. Without directly referencing Beuys, Schechner (2003, 29) notes in this context that "every acorn is an oak-in-process. But between acorn and oak is sun, rain, wind, lightning, and men with axes." Weather, objects, and human interference are included in equal measure in these processes, and they are ready to be actualized. Clouds, rain, and temperature are active contributors to any such event as it evolves. Transformation in an actual is neither reversible nor detached from its surroundings. It cannot be revised or corrected but all conditions become part of the piece. After all, what would Woodstock have been like without the rain? What would Burning Man be without the heat of the desert? And we might add: What would a performance including *Tweenbot* be without the heat of the summer day in Washington Square Park?

Peters (2015, 2) argues for "existential media," wherein media are "vessels and environments, containers of possibility that anchor our existence and make what we are doing possible," defined as "ensembles of natural element and human craft" (3). Peters turns away from media as meaning-carrying channels and argues, for example, for clouds to be media. Vital media do not follow this argument to declare water vapor itself a media, but it certainly agrees with the idea that all components are necessary to coproduce the event. All are active, all transform together. This means that any concept of media as channels dissolves. Media, we recall, are networks of activity. These networks only manifest through the activity, which can be designed but is ultimately emerging from the situation. As they come to live, their frame expands, and all partners involved change and grow. That is why it is too limited to identify transformations only among humans. Vital media are not clouds (unless a stunt pilot writes patterns into the sky), but they are irreversible shared actuals that transform everyone and everything involved. These changes might indeed include water.

Describing the actual, Schechner discusses Kaprow's *Fluids* (1967), a piece that included ice structures constructed of hundreds of ice blocks and positioned in the cityscape of Los Angeles, where they were left to melt. Melting is the active contribution of the ice blocks assembled by the human constructors. The event unfolded within the context of the urban landscape, in the Californian weather conditions, encountered by audiences passing the mysterious structures. The temperature of Southern California, the physical capabilities of ice (e.g., binding the blocks using salt), the urban settings that housed the structures (e.g., the flatness of a surface), the navigation of builders and audiences—all shape the *actual* in Schechner's sense. They form a network that actualizes the happening of *Fluids*, an irreversible, interdependent, yet clearly defined change. The emerging combinations of all partners involved provide a new progression, a novel condition, a happening. This relates to Austin's notion of "performativity" but in a material dialogue.

For J. L. Austin (1962, 6), starting from a focus on verbal communication, "performative" refers not to a description or reflection but "to do it." One example for such a defining "doing" is the declaration of "I do" during a traditional Christian-style marriage ceremony. "When I say, before the registrar or altar, &c., 'I do,' I am not reporting on a marriage: I am indulging in it" (1962, 6). The words themselves cause irrevocable changes to the status of the participants. They are examples for an *actual* through a performative action that defines not only the life of the couple (who utter the vows) but also of the assembly (who witness the event), the role of the ceremonial space (like a church), any symbols involved (like rings or dresses), and their acceptance and legal status in society (like certificates and changing tax status). This realization constitutes a definite change, and the materials involved are not mere representations of a human change, but they are an integral part of the complete activity. We see this in our tribal rites, the encounter with the gun in *Rhythm 0*, the melting ice encounter in *Fluids*, the growth of the trees in *7000 Oaks*, as much as in the destruction of *hitchBOT* or the progress through a video game, such as *Mad Dog McCree*. The changes might be differently weighted but a performative alteration is taking place, and we can measure the quality of the design at hand by how effective it is being supported. Austin centers foremost on the human production process but the uttered "I do" has legal ramifications; it reflects on the role of the space that hosts and coshapes the utterance; it forms a community in the here and now of the ceremony itself—however momentary that assembly might be. Far beyond the social, it shapes the balance between the cognitive and noncognizant participants.

Vital media, from such digital formats as video games, online worlds, and internet-of-things applications to robots, performance props, and other material encounters, operate in this logic. They see performative action as constitutive while engaging the human participant as one ingredient among many in the unfolding events. Their transformations are instances of actuals supported by technology, collaboratively enacted by at least one person and many objects within context. For better or worse, playing *Mad Dog McCree* can be a transformative social performance, just like a marriage ceremony or a rite of passage into adulthood. The activity is not limited to a "looking back" to rediscover the cultural history of the tribe or a reenactment of personal memory. It is a forward-looking production and a creative act emerging from freshly formed connections in the moment and through transformations of the active partners. It positions audiences and performer as much as it realizes materials, objects, and media. We will later see this applied to crafters as well as to performers.

Vital media actuals like Kaprow's *Fluids* become part of a larger-than-social sculpture because they form one's dialogue with others as well as with things and materials. It is the task of vital media to encourage this dialogue, position the participant in it, and facilitate this form of intra-actualization. Their focus is not on the realization of an internal shift, a self-perceived transformation or exploration of a "second self" (Turkle 1984). Instead, these are world-building exercises that form through mutual dependencies. As these shared instances are produced, such cognitive states as flow or presence can be effects, but they are not central to the design of the immediate encounter, which does not circle around the user experience alone anymore. What counts are the manifesting bonds between the surroundings, material and immaterial, cognizant and noncognizant, analog and digital, nonhuman and human participants. Vital media are not human centered anymore but inclusive and distributed.

This is where we find the role of the human participant, not as a dependent or master, but as collaborator in an increasingly decentered network of producers where the feedback loop is extended to all participants. This puts an end to the idea of the "user" as well as to "human-centered" designs that target a transformation of the interactor. It replaces them with mediated actuals that transform all intra-acting partners. Ultimately, the question of acting in digital media is answered via three key points: (1) the activation of objects, by emphasizing their material agency; (2) the positioning of the human participant, nonhierarchically as collaborator and contributor of their willed expression; and (3) the support of the emerging actual, which relies on the agency of the noncognizant as much as that of the cognizant, and can lead to a transformation of all partners involved.

Performing Media

We have encountered an argument that expands performance to a wider action network. One that has gradually chipped away on a humancentric view and filled in the material agencies of objects, including digital media objects. The following section reverses the perspective and argues from digital media toward performance. If the sections above tackled the role of the human and materials in the performative creation of vital media, then the following ones investigate the role of the media technologies. This includes brief revisits to key points made earlier but encounters them from a different vantage point.

To describe the particular media format of video games and players' encounters with them, Espen Aarseth (1997, 1) has suggested a textual machine that requires human participants to physically tackle digital media: "During the cybertextual process, the user will have effectuated a semiotic sequence, and this selective movement is a work of physical construction that the various concepts of 'reading' do not account for." This *ergodic* component, such as pressing a button on a keyboard, moving a mouse controller, or using motion sensors, is central to Aarseth's approach to the production of text in video games and digital media at large. His "textual machine" describes the moment of shared production laid out above. Neither human nor object/media can perform independently of each other; the text of the video game only comes into being through collaboration when all partners contribute to the productive action. The human contributes ergodic action. A game might contribute its own action driven by AI, simulated physics, prescribed moments, or whatever form of simulation. The platform running this code might contribute frame rate, physical controllers, and various forms of interaction design. For Aarseth, the activation and operation of this machine is key, and he worries much less about questions of creative expression. Consequently, when it comes to vital media, the focus on the ergodic is a good indicator of the human engagement, but it can also be misleading. An ergodic action that lacks the notion of willed expression might be an activity, but for vital media it would not necessarily qualify as a media action. Breathing, walking, and carrying groceries are ergodic actions but rarely performative ones. In these instances, something human *does*, but it does *not* express. The production of human action does not conflate with the production of intended expression. Yet, as argued above, this expression cannot originate from the other partners. For vital media, the unique human contribution is not the muscle that presses a button but the cognitive reason for triggering that muscle. Doing alone is not performing. The necessary qualities of any action to make it willed expression depend on perception, interpretation, reflection, and intention. They hinge on cognitive abilities and the realization of these abilities. It is the cognitive actor's unique contribution just as the rendered image of a video game is the computer's contribution. Aarseth rightfully located the ergodic muscular engagement as a necessary and identifiable contribution of the player to an unfolding game-play event. To this, vital media have to add the willed nature of that action. Muscles alone do not provide the necessary buy-in. Other participants might contribute their own unique abilities and necessities to produce action, but this should not be misinterpreted as performance in the case of vital media. Video games might feature AI bots that play each other in a video game; learning robots might develop elaborate behaviors and emergent behavior. Without cognition and intended expression, this remains action without becoming performance.

Proponents of advanced artificial intelligences would take issue with this position, but this argument is not aimed at them. The main question for vital media remains tied to the human endeavor. If one attributes to AI the need to self-construct and thus self-express, then one veers too far from that target. Staying committed to the concept of vital media requires digital media to concretize human positioning in the world and vice versa, not to replace it. Neither does it allow us to reduce the human body to its material functions to level it with surrounding materials. The differences remain, but the challenge is to activate them in creative collaborations.

Making this connection concrete also means that the prior dominance of the human action has been dispersed; it has become but one component in the unfolding act. To recapture the evolutionary thread introduced by

Welsch: performance is not a rising above the material or a control through some higher functions. Instead, it is an addition to the already ongoing development of action. We, as human contributors, cannot generate "intended expression" without the external contributions from objects, dead or alive, virtual or physical. Here we find the role of materials as well as digital technologies in the performative contributions. This is the "machine" of vital media, which is necessarily distributed, emergent, nonconclusive networked activity. In this case, performance is a productive act. It is not an analytical perspective.

Schechner (2003) would largely argue for a humancentric idea of performance, but he suggests that even the collaboration of two machines might be interpreted "as" performance, as long as it can be framed as such. Discussing activity "as" performance uses performative means to analyze an encounter. It puts the event into a critical frame and uses elements of performance for that critique. If two robots meet, did they appear friendly or hostile? Can I interpret their activities in a certain way? Can I identify such elements as a beginning, climax, and end in the encounter? In that way, performance can be used as a critical perspective in relation to actions that were never intended to produce such a reflection. The machines, technical objects, and materials might provide actions that stimulate human interpretations. Because they can be anthropomorphized, they can easily be discussed as performances. Yet they remain a step detached from the original production that is so central here. To elaborate the differences and the role of code in such a performance, we remain in the world of video games.

For the procedural art project *ioq3aPaint*, Julian Oliver used a hacked—and ultimately open-sourced version—of the first-person shooter game *Quake III: Arena* (Devine, Carmack, and Cash 1999). Usually, hackers change code in games to achieve some game-play benefit, such as activating a cheat or breaking a copy protection. Oliver's hack was an artistic one. More specifically, he altered the way the images were rendered on-screen. Instead of deleting older rendered visuals, this version of the game merely painted over already existing ones. In this hack, the game engine does not properly "clean up" the screen between frames but instead overwrites it, allowing all moving objects to drag their textures across the screen as they move. This results in visuals that bleed into each other. Tracks of moving game objects produce lines of distorted movement as a result of the redraw glitch as seen in figure 3.3. The hack of *ioq3aPaint* alters the basic image-production

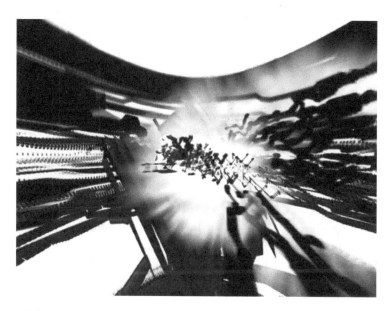

Figure 3.3
ioq3aPaint (2007). Used with permission by Julian Oliver.

process at that heart of real-time graphics of video games. This alteration makes the game unplayable for humans. They cannot read, compute, and react reasonably to the altered image world. But the remaining game logic stays intact under all the visual distortion, which means that nonhuman players, such as AI bots, can still interact with this world. They can still identify space, action, and the activity of other bots in the level as their analysis of the situation does not depend on visuals. Humans cannot play this game—but bots can. Utilizing this feature, Oliver started the game by pitting AI bots against each other and recording the results. The results are hypnotic painting events of AI characters battling each other in a moving image of blurring colors and textures. The game characters might be following their battle behaviors, but thanks to Oliver's hack they operate like AI-driven paintbrushes inside the 3D game level.

The project *ioq3aPaint* only works because Oliver intentionally altered the game engine. But the unfolding event itself does not focus on any further human interaction, and the game is literally rendered unusable for human players. This exclusion of the player is part of the piece's nature. It certainly can be interpreted "as" performance: the painting unfolds dramatically as the AI beings battle each other on the stage of the game world. We can see visual dynamics unfolding and cryptic spatial relationships. But this does not qualify as a performative production moment in vital media. *ioq3aPaint* does not require any human enaction itself. Humans are reduced to interpreters or prescribers. This makes it clear that digital technologies, objects, and materials cannot perform by themselves.

For our purposes, such machine actions become interesting when they support relations across domains, highlight interdependencies, and encourage contributions from their human as well as nonhuman contributors. They are powerful vital media facilitators when they bring forth nonhuman and human qualities. The results might be destructive, as in the demise of *hitchBOT*, or supportive, as in the help provided to *Tweenbot*, but they are always fundamental to the emerging event and cannot be reduced to mere interpretation, as in the visuals of *ioq3aPaint*. In vital media any interaction is a concrete and irreversible process that manifests a specific condition through enacted willed expression and material action. They do not include *ioq3aPaint*, which does not depend on human ergodic contributions but by design excludes such intervention. This does not take anything away from the piece's artistic value or its innovative design but helps us to differentiate vital media from strands in procedural art generation or AI-based pieces.

The past sections have connected material and human performance components to media and the actualization of events. Things—noncognitive entities—have been identified as able to act but incapable to perform themselves. Yet they remain central to vital media as no performance is possible without them as participants. The position of the human participant as contributor of the expression and cocreative partner in the process has been laid out. In this case, expressions are outpourings of cognitive engagement. They are based on necessities of cognitive development and thus a "need" of humans. They help the human contributor to constitute one's own position and identity, while the material agencies help the objects to further their qualities. This positions the agency of materials and that of humans in a forever-connected feedback loop, a material version of Fischer-Lichte's autopoietic one. We are in an interdependent flow state, where the partners coconstitute each other. Neither can emerge alone. Finally, the role of digital systems (like game engines) was discussed and lines were drawn between actions that are supporting vital media network and those that are not.

But this also leaves a number of questions open regarding this collaboration. How does this emergence happen in practice? Are there existing forms within performance that can show us these interdependencies at work between human and material? Tackling this question we turn to puppetry and material performance to identify examples of the interplay of active and performing entities, cognitive and noncognitive beings, and their inherent interdependencies.

Puppets and Material Performance

Puppets exist on the threshold of human and nonhuman encounters. "Whenever someone endows an inanimate object with life force and casts it in a scenario, a puppet is born" (Blumenthal 2005, 11). The "inanimate" materiality of the puppet is a defining component as much as the "life force" in the emerging performance. Puppetry is inherently a vital media practice because it only emerges from a cocreation that uses the differences and the synergies between the puppeteer's and the puppet's bodies (Jurkowski 1990).

Heinrich von Kleist ([1811] 1982) praised the benefits of mechanical puppets over even the best-trained actors precisely because they lack consciousness and thus can act freely and beautifully in ways inaccessible to humans, who are limited by their self-awareness. As helpful as his polemic is to clarify the importance of the material in the production, we have to avoid any such weighing of one partner against the other, which only rebuilds hierarchies. Puppets are relevant here not because they may replace humans but because they are a format that inherently emphasizes the dependencies.

It might be more productive to look at puppets as an example of Haraway's idea of *sf*, or string figures. String figures are part of Haraway's argument for the kinship between humans, critters, materials, and any other object. They are a methodological tracing of connections as much as a look at actual objects and the practice of shared emergence (Haraway 2016, 3). String figures are dynamic connections, fluent networks in which humans have to realize their kinship with all other partners. In puppetry this manifests in the relationship of puppet and puppeteer. A shadow puppet, glove puppet, and a marionette become—in some cases, literally—string figures when they are played. Dassia Posner, Claudia Orenstein, and John Bell

(2014, 5) define puppetry as "the human infusion of independent life into lifeless, but not agentless, objects in performance." Puppetry is not only an example of string figures in action; it is also exemplary of the emergent and interdependent qualities of the vital. But what is "independent life" in this definition?

Popular answers in puppetry often lean on forms of animism and anthropomorphism. Audiences might project life into a nonliving thing by projecting specific meaning into it, by interpreting it. One might see an archetypical family in the assembly of particular geometric shapes or project character traits into the jerky movements of a found object. Such an infusion depends on audience interpretation. It might be part of the perception- and cultural meaning-making efforts, but a production-centered approach focuses on different mechanisms. If we trust in the agency of materials, then we should realize that they do not require anthropomorphism to become actors. A puppet's actionability is not produced by human interpretation, but it is inherent in that object and its materials. The first thing many puppeteers do when they pick up a puppet is to look at it to let it speak to them on its own terms. The puppeteer listens to the puppet to establish a mutual connection. This moment happens in an intimate encounter with the material. Any nonliving agent involved—metal, wood, leather, felt, fabrics—works and contributes to the overall outcome not because of some projection but because of its active contributions that depend on its specific fabric, leather, wood, felt, or metal qualities. They produce particular actions because they are of their very own nature. Every puppet differs in its expressions because it offers unique forms of those material qualities. Kermit the Frog is a hand-and-rod puppet, who has difficulty walking as his legs are merely dangling off his felt body. A Harlequin marionette offers control over all limbs but, due to its fixed wooden head, lacks the facial expressions that a felt hand puppet has. This "lifeless, but not agentless," activity is foundational in any puppet performance, but it needs a human participant to become vital media.

The human contributions to puppetry remain a form of constant relating to the material object to produce expression with an intent. It is one action to reach into a sock to turn it inside out before throwing it into the laundry basket. It is another to reach into it and manipulate it while voicing a monologue on the traumatizing perils our fabrics encounter in washing machines. The object is a relational and cooperating thing through its material, its operational properties, its own network of contexts and interconnections. This sock not only speaks through the movement of the hand within it but also through the fuzziness of the material, the color, its thickness, texture, the holes that earlier events or moths might have ripped open, maybe even the smell. It is important to realize that the material works *because* it is different from the human. Puppets and puppeteers connect in an ambiguous copresence as "two beings who belong to two different modes of existence and actuality" (Piris 2014, 41). Their collaboration can only evolve in the distinction of their modes. In that way, puppetry exemplifies the necessarily different yet codependent creative development of an expressive action in vital media practice. Only because of their differences, because the puppet has unexpected qualities and features that can reveal themselves in the encounter and inspire the puppeteer, can the puppet and the puppeteer meet and perform together. The qualities of the puppet interoperate with those of the puppeteer to collaborate toward a shared event. That event is the puppet performance, and it is a moment of vital media production.

As noted above, Kleist ([1811] 1982) argues in his treatise on marionettes that humans' consciousness is in the way of their reaching the grace that lifeless marionettes are capable of. When we are consciously performing a dance, he argues, we distort and contort our bodies to produce certain moves. We attempt to replicate moments of surprising beauty but fail because we are aware of them and try too hard. Being self-conscious and operating through the sheer force of will makes natural movements difficult. "Such blunders . . . are unavoidable, since we have eaten of the tree of knowledge. But Paradise is locked and bolted and the Cherub is behind us. We must make a journey around the world, to see if a back door has perhaps been left open" (Kleist [1811] 1982, 3). In comparison, the noncognizant wood puppet might perform a movement with ease. It does not know self or shame or pride or any cultural limitation but only follows the rules of gravity. Puppets are the way of the back door because they remain untouched by consciousness. Their value is their lack of sentience, their detachment from worries about paradise.

Kleist suggests the superiority of the puppet as a stimulating provocation. To provoke and highlight the object's powers in performance, his treatise shifts the focus too much onto the material side. He acknowledges the role of the puppeteer but emphasizes the impact of mechanical forces over it. The center of gravity and the motions of an object in relation to it outshine the puppeteer's controls in Kleist's argument. This pushes the conscious human out of the equation entirely, which is why any design following his argument would fail as vital media. The balance tips too far to the material side. It is the encounter of the puppet forces with the human ones that has to remain central, and they not only allow the material to act but also the human to engage and expand their cognition.

In this encounter, puppets offer "concrete means of playing with new embodiments of humanity. To understand our engagement with puppetry is to chart and reveal new expressions of ourselves" (Posner, Orenstein, and Bell 2014, 2). New expressions are brought forward through this collaboration, through the activation of the puppet's mechanical qualities as much as the human's handling. The result has been described as "material performance": "At its simplest, this term assumes that puppets and other material objects in performance bear visual and kinetic meanings that operate independently of whatever meanings we may inscribe upon them in performance" (Posner, Orenstein, and Bell 2014, 5). In the operation of material performance, the puppeteer's and the material's attributes meet and manifest in action. The qualities of wood are still part of the way a carved marionette works. Its weight, gradual deformation, and rigidity affect the physical movements of the puppet, and they follow Kleist's argument as they are in conversation with the work of the puppeteer. Namely the puppeteer's controls are relating to these features, the weight, the joints, or the shadow a puppet might cast.

Trixie la Brique (pictured in figure 3.4) is a standard-sized construction brick. She is also a puppet performing in *The Brick Bros. Circus* by The Puppetmongers and on stage in figure 3.5. *The Brick Bros. Circus* is a found-object puppetry show starring almost exclusively bricks. Some bricks might be dressed up (Trixie wears a little tutu); some remain bare. Their "brickness" is an integral part of the performance. Their weight, shape, and even their rigid unresponsiveness are part of the play. During the show, brick puppets sometimes seem to refuse prompts provided by the two human puppeteers, Ann Powell and David Powell, by stubbornly denying actions, by being bricks. At other times, they might break role and engage, even reveal different natures. One puppet turns out to be a foam-built brick imitation that can be squeezed and deformed. Its material difference to the other puppets serves as material revelation. The nature of the brick codefines the puppets'

Figure 3.4
Trixie la Brique as preserved at the archive at the Center for Puppetry Arts. Used with permission by the Center for Puppetry Arts.

agencies and their actions. Trixie, for one, goes on to perform a tightrope balancing act.

It is the encounter between the two inherently different entities that is part of the attraction: the seemingly rigid object puppet and the puppeteer who builds novel relations to it and together they generate new contexts. They team up to create a puppetry show. But it is not only the puppet that is recontextualized here. A puppet encounter like this is destabilizing and

Figure 3.5
Ann and David Powell performing with Trixie in *The Brick Bros. Circus*, by The Puppetmongers. Used with permission by Ann and David Powell.

questions our own role as well, as John Bell (2014, 50; emphasis in original) argues:

> Modern puppet performances can be threatening, doubt-inducing, and anxiety-provoking events because they remind us that we are not necessarily in control of as much as we thought we were. Modernity has traditionally asserted its confidence in human potential, in our rational minds, in our ability to impose logic over untamed and illogical features of our world, including societies we consider un-modern, and even over nature itself—the ultimate force in need of taming. But play with puppets, machines, projected images, and other objects is constantly unsettling because it always leads to doubt about our mastery of the material world. If that piece of wood, that lump of clay, that shadow figure, that machine, seems to be moving of its own accord, then where are we as humans? The essence of puppet, mask, and object performance (as countless puppeteers have said from their own experience) is not mastery of the material world but a constant negotiation back and forth with it. Puppet performance reveals to us that the results of those negotiations are not at all preordained and that human superiority over the material world is not something to count on, especially since *we* all eventually end up as lifeless objects.

The encounter between human and material can be reflective and critical. It can question human qualities through the actions of nonhuman contributors. It interrogates conditions between them. Such an encounter with the material is never complete but constantly broken, questioned, dialectic, always in conversation, never secure in any assumption of the material's contribution or our own. This challenge has been discussed in puppetry by referencing Brecht's *Verfremdungseffekt* (alienation effect) (e.g., Jurkowski 2014). The Verfremdungseffekt is a dual-edged sword itself. It avoids emotional identification with a character onstage and rejects a cathartic resolution, but it supports through that distancing a critical perspective and ongoing criticism. It keeps the dialogue open during the performance and invites critical reflection on the unfolding action. In puppetry, material puppet and human might collaborate, but their collaboration also reveals critical differences and remains dialectic. We constantly play with the different agencies without ever resolving them. The event is constantly emerging through collaboration, constantly at risk, and might break any moment. Trixie la Brique and her puppeteers do indeed perform on a tightrope.

It is a defining element of vital media to realize this condition. They facilitate contributions from material and human contributors arranged in a way so that both can be activated, questioned, and brought forward. To support this, the productive performative moment has to become a critical collaboration. Its design cannot presume hierarchies, feed resolving pathways or produce closure, but it needs to emphasize collaborative critical exploration. Vital media do not preimpose a solution or outcome but support this kind of probing encounter. As all acting partners are engaged, they are also questioned. The results are "not at all preordained" but emergent and destabilizing. They are in need of continuous rebalancing.

The focus on activity and production means that vital media are married to an eternal construction process. They never resolve, never conclude, continuously producing new contributions and questions to this dialogue. This happens in *The Brick Bros. Circus* performance when a brick starts to dance on a tightrope. It happens every time when we engage with vital media and encounter a potentially "threatening, doubt-inducing, and anxiety-provoking" collaboration. The driving question for vital media is how to design activity networks that facilitate this dialectic production process itself and to avoid conclusions but support sustainable continuations that foster critical engagement.

Carl DiSalvo's (2012) notion of "adversarial design" is rooted in design studies but relates to these challenges of continuous performance and production. His approach builds on the concept of "agonism," a condition of disagreement. It persists through dissensus and confrontation. Adversarial design aims to support agonism by providing means to produce within a state of dissensus. No ultimate solution is targeted or expected, as the goal is not a final state but a continuation of that dialogue. "Agonism is a condition of forever looping contestation. The ongoing disagreement and confrontation are not detrimental to the endeavor of democracy but are productive of the democratic condition. Through contentious affect and expression, democracy is instantiated and expressed. From an agonistic perspective, democracy is a situation in which the facts, beliefs, and practices of a society are forever examined and challenged" (DiSalvo 2012, 5).

Where adversarial design targets ongoing critical examination of our political conditions, vital media arrive at a comparable critical interrogation of the emerging material ones. It tackles design in collaboration with materials, not as a sociopolitical necessity but as an evolutionary one. But these dynamic coproductions among humans and nonhumans need to be constantly challenged, just as requested by adversarial design.

For DiSalvo, *Blendie* shows the tension between human operator and robot, and it is an example of the uncanny relationship caused by a machinecentric embodiment. It is the human body that has to follow the robotic one, not the other way around. This shift changes our perception of the human-robot encounter. "The tension that is identified and brought to the fore concerns how we conceptualize what is human and what is machine and how these conceptualizations inform and interact with each other" (DiSalvo 2012, 71). This corresponds to the idea of performance as a coproductive act, as mutual conceptualizations. We define our own identity in relation and through shared activity with that around us. As in vital media, the differences of entities are not ignored or blurred out, but they are part of the tension within the encounter.

DiSalvo's reading eventually turns to the example of social robots and how material encounters—such as screaming at a blender robot—might manifest social organizations. For DiSalvo, the results reflect political conditions. They "can be manifestations—expressive encapsulations—of some aspect of the political condition" (DiSalvo 2012, 117). Adversarial design looks at such political form and dialogue. Vital media look at our being with material, which ultimately affects the political. But vital media predates social impact. That is not to say that there is no such political context.

Puppetry has a long tradition of subversive techniques that have responded to political conditions. Various puppetry forms have been censored by state, religious, and other social institutions over time. The representation of animals, humans, or gods caused tensions in the Wayang puppetry formats.

The often anarchic nature of Judy and Punch or Kasperl performances explicitly undermined social conventions in Western Europe. Ventriloquists have made an art form of dramatizing the insults of the "other" puppet that technically speaks with their voice. Even mediated, puppets remain spiteful, as seen in TV puppetry shows such as *Spitting Image* (ITV 1984–1996). The puppeteer could hide behind the material object, which would speak truth to power without the fear of reprimands. The resilience of puppetry in the face of censorship and prosecution, against state-sponsored expression, has led to a form of folk art that emerges less from institutional support, like opera, but operates from the ground up in hard-to-regulate conditions. It remains largely unconcerned about distinctions between high and low art but keeps an eternally critical eye to the world it emerges from and in. The argument returns to this folk-art relation in the closing of this book, but it might serve as a pointer to the way the material dialogue ultimately reflects the political one.

Material performance in puppetry turns critical, both as a sociocultural statement within the communities and as an encounter of humans with materials. Leather, fabric, wood, strings, rods, paper, and fish bones all contribute and help initiate this dialogue. A reversed perspective to *Blendie* might be that of the ventriloquist: if *Blendie* is machine embodiment that requires the human to adjust to the operation of the object, then a ventriloquist operates through their body. They cultivate the ability to produce and "throw voice," and the material puppet relates to this through its activities. In both cases, it is the splitting of identities and agencies, the uncanny shift, that helps to bring the performative event to the forefront. Out of this, an adversarial practice can emerge, but more fundamentally, it shares the call for an unending and unresolvable critical dialogue. Turning from this short reflection on puppetry to digital performance, the following example will lay out some shared moments of production and encounters in a different form of embodied digitally augmented performance.

Subway

I had met Molly Kleiman at a workshop organized by Eyebeam, an art and technology community in New York. The workshop was led by the performance studies scholar Richard Schechner and intended as an intercultural dialogue. A video link connected Iranian students and performers with a group of interested participants in New York to facilitate a discussion. During that discussion, we touched on differing practices, possible overlaps, and future projects. We started to develop a dialogue. In my case, this dialogue was cut short as I had to return after the first day of the workshop to Atlanta, where I teach at the Georgia Institute of Technology. But I stayed in touch with two co-organizers of the workshop, Kleiman and her collaborator Ava Ansari, an Iranian-born artist living in New York. Ansari studied arts politics at the New York University Tisch graduate program, where Schechner served as professor emeritus. During that time, Ansari and Kleiman curated an art space project, *The Back Room* (2010–2014), dedicated to supporting artistic dialogue where physical connections and exhibitions were difficult. The term "back room" refers to Ansari's experience in her hometown, Tehran, where cultural activities often happen in the back room of buildings to escape surveillance and censorship: "We wanted to take this metaphor of the back room and open it up. We asked ourselves: How can we help develop an expanded, immaterial 'back room'? How could such a backchannel provide space for a workshop? An exhibition? Technology can partially help us to manifest this. The features and limitations of this technology inevitably influence—and can even be the material of—artistic production" (Molly Kleiman in Ansari, Farzin, and Kleiman 2012). *The Back Room* saw itself as an underground pedagogical and curatorial project to blur boundaries and connect creatives. The Eyebeam workshop was one of their practical efforts, designed to connect artists across cultural and political divides.

When I started to organize my graduate class for the next term and was looking for collaborators to inspire new class projects, I reached out to Ansari. She became a collaborating artist for the course, and we organized a video conference with her to present her work to the students. The setup was not unlike the first workshop at Eyebeam but now connecting New York to Atlanta. In class, Ansari showed a range of her work and explained her underlying motivation and their context to the students. One piece presented to the class was *Dancing by Myself in Public*, a video piece filmed by Jian Yi in 2011 consisting mainly of a single-take video recorded in the Times Square/42nd Street subway station. As seen in figure 3.6, the video centers on a bench in the subway station and is shot from across a parallel track. The Times Square subway station is one of the busiest stations in the New York metro system and numerous passengers are caught in the action.

Figure 3.6
Still shot from *Dancing by Myself in Public*. Used with permission by Ansari.

Bystanders wait for trains to arrive, cross in front of the camera on the adjacent track; trains pass, stop, leave the station. The initial impression is one of the daily routine at a quintessential New York subway station. Ansari enters but initially appears to be just one additional character in this routine. She sits on the bench, waits, stretches, and gradually her moves begin to evolve into a dance. The dance becomes more and more agitated while obviously nondiegetic music (Mathew Jonson's "Symphony for the Apocalypse" and Goldfrapp's "Strict Machine") plays on the video's soundtrack over the bustling sounds of the station. Bystanders look at Ansari performing her personal dance with mild interest but without much further concern. Eventually, the dance subsumes; Ansari sits down and strikes up a conversation with a man who happens to be sitting on the bench as the music changes to Schubert's "Ave Maria" (sung by Barbara Bonney). A train obstructs the view one final time. When the train leaves the station, Ansari is gone, leaving the shot to linger on the bench and the remaining commuters who still go about their own ways.

When Ansari presented the *Dancing by Myself in Public* piece to the class, she mentioned that the project was originally driven by her hope to share the dance with fellow Iranians, who lacked this freedom of expression due to restrictive rules in Iran that punish public dancing. Dancing in public is heavily restricted in Iran and severely punished as "indecent" behavior.

One of the graduate students, Andrew Quitmeyer, picked this piece as the reference for a digital intervention he built in response. Inspired by Ansari's original statement, the question for the digital intervention was how to bring the dance practice to Tehran without directly breaking any of the draconian laws in effect in Iran and endangering the safety of any potential participant. To achieve this, the new version revised and reframed the notion of dancing itself. It did so by using vital media practices.

Ansari's original piece was clearly marked as a video performance. It played with unwitting participants strolling into the shot, subway trains rushing by, but the role of the performing dancer, Ansari, and that of the videographer, Yi, defined in the piece. It was framed as a visual piece, had a clear beginning and end, which were supported by non-diegetic use of musical cues and was driven by a central intent. In Quitmeyer's adaptation, many of these given conditions opened up, moving the new version closer to a vital media piece.

Technically, Quitmeyer took about one thousand continuous frames of the video and traced the contours of the dancer in a postproduction video manipulation program (After Effects). These images, then, served as inlays for an augmented reality (AR) mobile app that he coded for the Android platform. Looking through the finder screen of their Android device, participants would see a still image of Ava's contours traced from one frame taken from *Dancing by Myself in Public* overlaid over the current live visuals in front of their device's camera (see figure 3.7). A second participant, then, could follow the guidance of the one operating the phone and recreate the pose in which Ansari's body was captured. This restaged pose could then be captured as a photo. In that way, pairs of participants could restage single dance poses one by one with the help of the app. Their individual still photographs would provide unique restaged frames of the video piece, each at

Figure 3.7
Image development from original video to posed shot in Iran. Used with permission by Quitmeyer/Ansari.

a new location. The mobile app was coded in Android, a platform chosen because of the widespread nature of this operating system but also because software distribution in the Android environment was easier compared to the more restrictive policies of the Apple app store system. At the time, Android allowed sharing and installation of external applications on local phones in a way that altogether sidestepped a single controlled distribution platform. This was a necessary feature as all participants needed to remain anonymous and digital traces had to be avoided, or at least hidden, to prevent possible conflicts with local censorship and prosecution.

Each installment on a local Android device provided access to a random selection of images taken from the roughly one thousand frames captured. Images did not follow any particular order but consisted of disjointed single takes. No two reference pictures were the same. Participants in Iran installed the app on their Android devices and started to experiment with the new piece following another online workshop we did between our group in Atlanta, Ansari in New York, and interested participants in a gallery in Tehran. Following this second workshop, a range of Iranian participants then restaged the individual images in public and private spaces in Iran. With the help of temporary email accounts, their images were collected by Quitmeyer in Atlanta.

The next step in the project dealt with the assembly of a second video piece that combined the original *Dancing by Myself in Public* video with the collected still images. Edited together in their numerical order, the still images taken in Iran produced a dance motion through the continuity of the moving bodies from one frame to the other. Even though the backgrounds change in every frame, the central movement of the dancing-posing bodies is continuous. The various performers' bodies followed the app's posing mechanism and thus stayed in the same visual space and scale. Arms, bodies, and legs might belong to different performers, but thanks to the digital app's operations they move in consistent directions across frames. The surrounding spaces turn into a flicker of single-frame backdrops, but the bodies and their positions in space ultimately provide a coherent dance motion. This motion is neither that of Ansari in *Dancing with Myself in Public* nor that of any single performer. It stems from many bodies and is reassembled as a visual effect every time the video is played. The dance only comes to life as a media object that is performed itself frame by frame. To guarantee the anonymity of every performer involved, a continuous black bar was added to the image to hide the performers' eyes and anonymize them in the overall assembly.

A second feature of the *Subway* app was that each participant could take images of their own poses and send them back to the team in the United States. These freestyle poses did not follow any guidance from Ansari and were entirely inspired by the Iranian participants. They were later used to inform a separate dance performance by Ansari. Her poses produced in New York had informed the Iranian performers. Now, the poses produced in Iran informed a choreography in the United States and instigated a two-way dialogue through the mediation.

Subway exemplifies the use of physical matter, digital media, human bodies, environments, and material context in a specific sociopolitical-material condition. It realizes the production of expression through the agency of all components. Willed expression remains central to the action, but identifying any single author, tool, or platform would be missing the point. Ansari's body dancing in the subway station in New York, the human bodies restaging individual frames of that dance in an incomparably more dangerous situation in Iran, Quitmeyer's dis- and reassembly of images in Atlanta—all these activities depended on cognitive engagement and the presence of a human body at that moment in that situation. At the same time, Android served as a participating platform shared among a great many devices, email accounts came and went, online video platforms facilitated meetings, and offline videos assembled a cacophony of images. The varied actions of digital components and material and human contributors added up to the production and the unfolding new dance event.

The separation of contours from Ansari's dance video, the operation and marketplace conditions of the Android app, the distribution of the digital images and apps, the final reassembly of images, and the real-time illusion of the dancing bodies are digital media operators during that dance production. *Subway*'s dancing as an expressive activity of the human body depends on a complex network of active ingredients and can only be realized through active mediation. This realization is denied any closure. The production emerges as a nonconclusive creative action that shows alternative conditions without resolving the tension. Bodies dance while the backgrounds flicker, reflecting the very condition the piece set out to engage with through an active and mediated production but showing the unresolved challenge as much as the constructed movement event.

Some parts of the digital components aged fast. The Android app is outdated by now, email accounts have been deleted, and digital traces fade not unlike they did in the virtual *7000 Oaks* piece. Indeed, their fading is part of the piece. The digital disappearance helps to keep participants anonymous and is part of *Subway*'s design. But the digital nature makes it also difficult to set a point of closure to the *Subway* piece. The images sent back to Ansari can inform potentially countless future dances, and the mediated dance happens every time the video is played, reassembled by the video player reading the file and producing the data stream for the human eye. Mediation, in this case, is a central form of ongoing production. This production happens in the rendering of the images, but it reaches across all aspects as the piece evolved. This included the chosen technological platform (Android), the communication channels (anonymous accounts), and the final aesthetics (anonymizing the participants), as well as the performance situations in Iran and the United States. In that way, *Subway* works as a digitally mediated social sculpture in which the digital is inherently included within a mixed network of human and nonhuman components.

Subway exemplifies many parts of a vital media production in the way it applies digital technology in the production, the performative expression constructed by individuals as well as the whole group, the embeddedness in a system of laws and restrictions, and in the way it responds to them. Like *7000 Oaks*, it does not resolve any of those tensions, but it continues to question the conditions. It might be credited to Ansari and Quitmeyer but manages to incorporate the digital media dimension into the unraveling conversation, which focuses much more on collaboration than individual authorship.

Leaving Performance

One could view performance as a quintessentially human activity, framed by cultural conditions and constructing social realities. Expanding these frames to include forms of production (McKenzie 2001), mediation (Auslander 2008), and material agency (Barad 2003) shifts the center away from such human dominance. Performance studies itself is debating these shifts.

Within this domain, puppetry is one area where these debates unfold, but there are other related practices that struggle with the same challenges.

Hans-Thies Lehmann (2006, 30–31) argues that modern theater turns altogether away from a dramatic form that once served the consolidation of "a cosmos whose closure was guaranteed." Instead, today "the mode of perception is shifting: a simultaneous and multi-perspectival form of perceiving is replacing the linear-successive" (Lehmann 2006, 16). For Lehmann, classic dramatic structures are not sufficient to feed these forms of perception. That is why he proposes the term *postdramatic* as a concept of performance that still recognizes past aesthetics but goes beyond drama and beyond such classical forms as constructions of fables or characters. This argument traces changes through many questions in performance, and Lehmann explores the idea with the help of numerous references. One that relates to the moment of material production is an increased reliance on physicality: "As postdramatic theatre moves away from a mental, intelligible structure towards the exposition of intense physicality, the *body is absolutized*. From now on, it seems, all social issues first have to pass through this needle's eye, they all have to adopt the form of a physical issue" (Lehmann 2006, 96; emphasis in the original). Lehmann speaks of the physicality of the "techno-body" (96). The body becomes a—possibly *the* most important—technological object of the performance action, which also includes the props, stage, situation, and other material conditions. In many performative situations, the boundaries of this body are explored and altered. The transformations Schechner uses to explain actuals include such changes. Bodies might get scarred or marked irrevocably as they make their activity space concrete and "organic" (Schechner 2003, 46). These changes affect human as much as nonhuman bodies. From its opening example of the *Lunar Orbiter 1*, this book has argued that media activity scars not only the human contributors but the nonhuman as well—if not more, given our lack of awareness. Following the call of vital media, one needs to recognize that such a self-realization and encounter with the body is only one part of the event. *Subway* is a network of activities in New York, Atlanta, and Tehran, as well as in digital apps, server farms, and media players. *7000 Oaks* is an ongoing city forestation project in Kassel involving performance actions in the 1980s as much as daily encounters with the still growing trees today. These interconnected webs reach across material, media, and human contributions and form the playground for vital media. Recognizing one's own body as material within them does not center the activities on that human self anymore, but it might be one powerful learning experience as we deal with these open systems and realize our own dependencies. It also allows us to pay our dues to our nonhuman partners.

To approach vital media, we defined performance not as an act of representation or reenactment but as the *action to produce expression with an intent.* As such, performative activity is not based on social communication but on the material encounter, and, thus, it deals with the "physical issue" of bodies affecting each other and forming anew, never recreating, always making a novel situation concrete. It is performative in the sense of a social constitutive insofar as it allows the human contributor expression. It is performative in the sense of material agency insofar as it depends on material encounters and builds on their unfolding changes and the objects' own emergence. We see this in Beuys's trees growing in Kassel as they continue an ongoing social sculpture, in a puppeteer-puppet relationship, in the demise of hitchhiking robots, in the image data that generates dancing. In all these cases, communication, interpretation, and sense making might be happening, but it is more important to first activate these relational intra-actions.

These relationships build on different abilities and needs that are realized through interdependent agencies in vital media practice. For the human in need of individuation, they provide opportunities for expression. For the material in need of modification, they provide physical change. Vital media do so, not for the benefit of any single one partner but for that of a balanced and critical production process that acknowledges individual qualities and contributions yet refutes single authorship or dominance of any individual partner involved.

We have to abandon ideas of humans as "users" and "human-centered" design and instead turn to physical encounters between human and nonhuman bodies and different-yet-partnering agencies. We have also lost claims of authorship and closure as we face operational openness and inconclusive emergence. The resulting vital media can never reach a stable equilibrium. Creating and supporting this balance inherently depends on a continuously unstable conditions maintained through an autopoietic feedback loop between all partners. Adversarial design tackles such an ever-critical approach for social conditions; vital media see it as a critical physical encountering in an evolutionary development first. Both have to remain unresolved by definition.

This section has tackled the role of performing humans, collaborating material agencies, and media as contributors in supporting such an unresolved yet performance-producing state. It showed ways in which this collaboration unfolds as a critical engagement and how they contribute to the balance required by vital media production. The next section flips the perspective from the challenge of producing performative expression to craft as the performance of production. It asks how we can explore this physical encounter through hands-on making and its conditions.

4 Recentering

This section turns to craft to focus on making and productive practices that highlight the personal encounter with varied materials, each of them having their own specific qualities and abilities. Ehren Tool's practice is deceptively simple: he makes cups, adorns them, and gives them away. One of these cups serves as the opening example in discussing his practice and history as it tells a multilayered story and offers an example for craft as a critical material encounter.

The critical turn to craft research avoids a revivalist stance. It approaches craft as a form of physical making and critical reflection. Craft scholars such as Peter Dormer (1997), Glenn Adamson (2007, 2013), and Edward Lucie-Smith (1981) assist in a definition of craft, which emerges as cocreative material practices that are based on needs and that further individuation.

Need—the immediate necessity of an acting partner—is tackled first. Differentiating craft from other domains Howard Risatti (2007) defines craft through a framework of need. I will adjust this idea to recognize that needs apply to materials as much as to humans before turning to the emerging crafted objects. Diversity stands out as a quality that includes valuable traces of a shared material culture. But objects as traces of human activity cannot be the end point of this craft exploration. Instead, I will return to making as a form of coemergence and shared becoming. Such a collaboration rejects conclusions and requires time and openness. We return to Simondon to discuss interdependent individuation, a key quality that craft can foster and that leads us to recognize the importance of surprise and improvisation.

Starting with Richard Sennett (2008) and Dormer, the next section discusses how these processes can remain critical and open-ended through the steps of encounter, exploration, and collaboration. These principles lead us to educational approaches and craft's relation to critical making.

Finally, the work of Amit Zoran serves as a concluding example of a critical hybrid craft practice. Zoran combines digital fabrication techniques with traditional craft methods to create objects with new qualities and stories. His "hybrid assemblage" pieces serve as a possible practice that recognizes the differences, combines them, yet refuses to resolve them.

Making Cups

Let us put a cup in your hand. When you turn the cup, you spot the rugged and slightly oversized imprint of a one-dollar bill on one side and the drawing of a disfigured couple on the other (see figure 4.1). The cup itself is uneven, seemingly created in a rush with little attention to the usual signs of craftsmanship. There are splashes of blue and red glaze on it. The walls are bumpy; the rim is not smooth; there is no clean foot. On its inside wall, one can see traces of the potter's hand where it pushed against the clay when pressing the dollar-bill stamp into the cup's outside wall. The cup is functional but it is not pretty. It might be uneven, but it still works as a container of liquid, fulfilling its basic function as a crafted piece. It might be unique, but it is not marked as an artwork and it is hard to think of it as a classic work of ceramic studio art, which often distinguishes itself through expert technique or design. This is one of the many thousands of cups produced by Ehren Tool in an ongoing pottery practice that started around 2001. The following pages will dive into the specifics of the one cup to reflect on the overall making practice and to offer a gateway into the discussion of craft as critical practice.

The drawing we saw on one side of the cup is a partial copy of *Prostitute and War Wounded (Two Victims of Capitalism)*, originally created by Otto Dix

Figure 4.1
Ehren Tool's cup front (left) and back (right); photographed by Michael Nitsche.

Figure 4.2
Otto Dix, *Prostitute and War Wounded (Two Victims of Capitalism)*. Used with permission by Artist Rights Society; © 2021 Artists Rights Society (ARS), New York/VG Bild-Kunst, Bonn.

as an ink-and-pencil drawing on cardboard (see figure 4.2). Dix created the work *Dirne und Kriegsverletzter* in 1923 with the memories of World War I still fresh in everybody's consciousness—his own as much as his audience's.

Dix had volunteered for the German army, served in a machine gun unit in France and Russia, been promoted to a low-ranking officer, and returned physically and mentally scarred. Near the end of World War II, he was enlisted again into the Volkssturm, returning once again to war. Dix survived World War II and continued to practice and exhibit in both parts of divided postwar Germany.

Dirne und Kriegsverletzter itself shows two scarred victims: on one side, a naked prostitute showing a syphilitic scar formed like a bullet hole; next to her, a soldier still wearing his hospital gown, his cheek deformed by a grotesque injury that splits up his face. The picture was originally published in 1923 in the satirical left-leaning German journal *Die Pleite*, founded by Wieland Herzfelde, his brother John Heartfield, and George Grosz. *Die Pleite* was already being published illegally by that time.

When they published Dix's painting, the editors added a new title that framed the introduction to the picture: *Zwei Opfer des Kapitalismus* (*Two Victims of Capitalism*), leading to the widely used title of the image today as *Dirne und Kriegsverletzter (Zwei Opfer des Kapitalismus)*, or *Prostitute and War Wounded (Two Victims of Capitalism)*. Today, the original is housed at the LWL-Museum für Kunst und Kultur Münster. *Prostitute and War Wounded (Two Victims of Capitalism)* is an uncompromisingly critical depiction of the effects of war, the direct and indirect wounds inflicted, the stigmata it causes.

The war and its effects can be traced across a range of his pieces, including his collection of fifty etchings titled *Der Krieg* (*The War*), published around the same time. Dix stated that for him "all art is exorcism" and that "painting is the effort to produce order; order in yourself. There is much chaos in me" (cited from Fox 2006, 256). *Prostitute and War Wounded (Two Victims of Capitalism)* presents one turn in this work, an attempt to engage with that "chaos" without resolution. The art critic Willi Wolfradt noted about Dix's work that it works against any comforting conclusion. His painting *The Trench* (1923) was "painted to make one puke, not to comfort. . . . It is simply unaesthetic!—and, of course, that is just what Dix is" (Crocket 1999, 96). Being "unaesthetic" is not an artistic flaw, but it is a critical stance. Adversarial discomfort is the effect intended, and it is successfully achieved by Dix. Because of this effect, Dix was among the first group of artists whose work was denounced as insulting as the Nazis took power. It did not fit into the coming regime and was banned as "degenerate art." Dix was stripped of his professor position at the Kunstakademie Dresden, but he remained in Germany.

Even before the Nazis took control, the visibility of war cripples in the public eye would become part of the political battles of the Weimar Republic. On the one hand, organized public rallies of war victims and wounded veterans were organized. On the other, the government attempted to reduce the visibility of those whose bodies literally embodied the terrors of war (Ziemann 2013). What shall be visible and imprinted on the national awareness and what shall remain in the shadows was a discussion then and remains one today.

Ehren Tool is a potter working for the University of California, Berkeley. Tool grew up in a family with a strong military background. This included family members serving in World War II and Vietnam. Tool himself followed the tradition and joined the Marine Corps. He participated in the first Iraq war and continued service in the corps at embassies after that. After his return, he used the support of the GI Bill to study. He first wanted to become an emergency medical technician, but eventually he turned to ceramics and ultimately graduated with a major in art from the University of Southern California. His war experiences remained powerful influences throughout these changes. The war in Iraq had left its marks, and these marks show in Tool's pottery as well as in his practice.

Tool works alone at a home studio in Berkeley producing cups in his time off from his regular work at UC Berkeley. Occasionally, he travels to present on his practice and to throw cups. One journey took him to Vietnam, where his father had fought during his military service. Another journey led him to the European battlefields of World War I, closer to Dix's past. Each time, materials and imagery change. When throwing in Vietnam, Tool turns to clay from local sources. When throwing in Europe, his selection of militaria and stamps includes European insignia. The cups emerge from the time and situation of their production. They build context with the spaces and times they emerge from. This includes those produced in the United States. During the presidency of Donald Trump, his figure was found on many cups, for example. Even if the context and location is less historically focused, Tool might create a frame to situate his work. During an art residency at the Portland Museum of Contemporary Craft, Tool built a bunker-like stack of clay sacks, only to gradually tear it down by using the clay for his cups. His is the practice of a production potter. Tool had to single-handedly throw hundreds and hundreds of cups to perform a tearing down of the bunker-style walls into cups. It comes as no surprise that Tool identifies himself as a "potter," not as a "ceramic artist."

Tool is deeply connected to the community of Marines and veterans, but at the same time, he remains critical of war and the wounds and injustices it inflicts. His cups tackle what he perceives as a lack of awareness about war and its effects. Making the scars and effects of war visible has become a major motivation for his work. Since 2001 he has thrown, customized, and distributed tens of thousands of individual cups. Each one is quickly thrown without much concern for the evenness of the wall, the shape of the lip, or the bottom rim. Once a cup is thrown, Tool uses it as a canvas to apply numerous stamps, images, splashes of color. The use of any supplemental ornament is a contested practice in craft (Adamson 2007) and in Tool's case, the technique itself grew from a rebellion against one of his ceramic teachers who had rejected the use of any such ornament. The use of multiple, often shocking, visuals on each cup might be atypical for most studio ceramics but became Tool's trademark form of expression. The cups oftentimes carry images of dead or dying soldiers, ammunition, as well as other evocative symbols, medals, weapons, and slogans that are not typical for the average household mug. Their construction, their reference points in the imagery, and even their distribution are part of Tool's larger project to make us aware of the terrors of war. The cup with the Dix image was handed to me when Tool visited Georgia Tech to give a guest talk. Tool does not sell his wares but gives them away freely to other veterans, widows, politicians, museum visitors, activists, and strangers he encounters. Giving them away continues the cups original mission. It opens up a dialogue around these objects without offering any resolution to the questions they pose. Each cup is uniquely handcrafted and is a gift that cannot be reduced to a simplified product or commodity. These objects and their making processes do not conform to an art market or studio ceramics model in any commercial way. Instead, they continue their work through these means of distribution. This current discussion of the cup he gifted to me is part of such a conversation.

Cups offer an intimate platform for Tool. They are useful and seemingly harmless. They draw little attention as art objects, yet they still attract enough to be personal objects. They are intimate, as they move "from my hand to your hand to your lips," as Tool noted in conversation (pers. conv., January 27, 2017). Cups are objects defined by a function and build on a most personal encounter, yet they do so in an inconspicuous way. They encourage a kind of looking based on rudimentary functions. Making one look at the horrors of war is a main driver for Tool's activities. "How do you get people to look? It is really a challenge to get people to look at anything. There is so much screen time and images of splash and people's attention spans are nothing. . . . At least with the cups, they have to finish their beverage" (Tool, pers. conv., January 27, 2017). We have to drink. We have to use cups. Thus, Tool hopes, we have to look. Cups as operational objects are working their own part in the mediation processes.

For Tool, making cups is as much a gesture of outreach and dialogue as his own version of what Dix saw as personal exorcism dealing with his own experiences. "Some of it is self-preservation. I had all these great and noble intentions joining the Marine Corps, and the gap between the stated goal and the outcome was vast and painful. It's got to be enough for me that they are just cups. It's got to be enough that I am just a potter. I am not an artist" (Tool, pers. conv., January 27, 2017). Each cup could be interpreted as a spiraling discourse between Tool, the material and its modification, the gifting of the cup to its new owner, and the continuation of the "looking" through the cup's use. As Tool notes himself, "It is cool seeing four hundred cups on the wall, but it is much cooler thinking about four hundred cups passing through time and space and picking up stories." Cups turn into small social sculptures that unfold their own life.

The turn to traditional production pottery is not a rejection of digital media as such but presents a focalization. Tool uses digital techniques himself. The wooden stamp used to print the dollar bill onto my gifted cup was laser engraved, and the Dix image was put on the cup via a decal that used digital printing techniques. In other cases, Tool might look for new imagery on the web, use digital video for an installation, or photography to document the work. The crafting process might not center on the digital, but it applies digital technology seamlessly. This work does not counter the digital age, it lives with it. In its everyday function, the simple cup actively stands in opposition to the overexposure and dynamic spectacle of digital media. It provides a continuity that stands against the multitude and short-lived hype.

This crafting process remains aware of its own qualities and identities. Tool describes his role as that of a potter, a maker of cups, not as an artist. But this includes an awareness of the scope of pottery. Tool is well aware that "the cups will be around for hundreds or thousands of years. The meaning of the cups will change." A fired piece of ceramics is extremely sturdy. A Tool cup can grow older even than any of Beuys's oaks, and it certainly will outlive any current computer operating system. In that way, the cup object is both humble and big. It is an everyday object for daily use but one that can operate longer than any single human live span. It provides for a lot of time during which "the meaning of the cups will change" as they keep on functioning and telling their stories.

The cup at hand opened up a range of questions regarding our encounters with craft as practice, crafted objects, and material encounters. Tool's pottery is part of an ongoing conversation; it is profoundly informed by challenges we ourselves face (or should face) as well as the functions and qualities of clay. It provides focalization through tools and craft activities, whether these are 3D printed stamps or the gradual deconstruction of bunkers of clay. His is a practice that unfolds over time, whether this refers to Tool's own making practice or the long-term function of clay in the hands of those who use his cups. It combines material with expression in an almost therapeutic crafting practice as well as in the daily operations of the cups. At its foundations, it builds on needs and function as the cups support a human need and expand their mission to make humans look from there. The cup made by Tool is a manifestation of personal practice, and we can read Tool's own history and his ongoing struggles into it. It is also a material conversation from the making itself to the definition of the object to the travels and histories of that cup in our hands and at our lips.

Turning to Craft

Craft opens up a rich field of references and connects media design to long-standing traditions and practices that already feature many vital qualities. Many of those practices grew out of a coexistence between materials, crafters, tools, and communities as they matured and emerged over long periods. These practices change constantly but retain connections to traditions, resources, and local cultures. This slower pace and evolving community indicate that the milieu that Simondon is searching for in the technological world seems a lot closer to us in craft traditions than in most digital development cycles. Crafters read materials, build and optimize tools, and they do so situated in communities of practice that are defined by not only social but also local and environmental conditions.

As promising as craft as a domain appears on first look, a turn to craft is also dangerous in that it might invite a sentimental and inherently uncritical revivalist position. There have been a number of "craft revivals" in the past that attempted to go back to a mythological "good life" rooted in a "better" past. Vital media cannot turn back. It is in the nature of the evolutionary perspective that there can be no turning back, no Luddite rejection of technology, no construction of a better coming world as a return to a past one. Turning to craft as a field itself must remain unsentimental, critical, and reflective. But one can still learn from the revivalists' arguments.

During one of those craft revivals in the early twentieth century, Lewis Mumford (1934, 345) turned to the role of the machine: "The machine was a direct expression of its own functions: the first cannon, the first crossbows, the first steam engines were all nakedly built for action. But once the primary problems of organization and operation had been solved, the human factor, which had been left out of the picture, needed somehow to be re-incorporated. The only precedent for this fuller integration of form came naturally from handicraft." In Mumford's case, craft is applied to address the encounter with technology as the "only precedent" to include the "human factor" in the development of advanced technologies. Other earlier revivals, driven by the likes of William Morris and Charles Ashbee, also turned to handicraft as the only past reference available. The colossal impact of the Industrial Revolution propelled them to call on craft as a return to an alternative way of live. There were many changes to respond to as industrialization reshaped society. Industrial mass production led to mass-produced goods and faster fabrication cycles supporting the needs of growing populations. Not two decades after Morris's death, it also supported the weapons of World War I, the first war in the industrial age producing new horrors on the battlefield as seen and shown by Otto Dix. It led to fundamental social changes, as work and living conditions changed rapidly, working environments imploded, established social traditions broke away, and new ones emerged. The revivalists' dream to return to craft has to be seen in the light of these tumultuous conditions.

Historical craft revivalists, like John Ruskin, turned to craft to reject contemporary technology and society and call for social change to counter the dissociation of people from their work. Where revivalists can overshoot their argument is in the view of craft as a solution in itself and as a "way back." But their emphasis on craft as a practice for personal and social development is still valuable. This is craft, not with a focus on the crafted object but on the process of crafting in a changed environment.

One of these changes is the spreading role of media through technologies. If the first machine was a "direct expression of its own function" and Mumford (1934, 356) looked into the technological as part of the "mode of life" from such function, then the troubles only grew exponentially with the turn to the digital. A computer's multimedia capabilities make it hard to confine it to any single function. The powers of the computer change how we apply, practice, or criticize craft. They change how we think about craft as a domain itself. The digital age brought new means of production to craft, introduced new tools, hybrid practices, and new materials. It changed the distribution of goods and knowledge through online services. It changed the notion of communities as well as practices among them. Historic demarcation lines blur in the way internet technology changes how crafters sell their goods or in the additions that new personal fabrication technologies provide to a crafter's production practices, as seen in the Tool cup. The Tool cup also stands for new challenges posed by a new culture of attention largely built on internet services. The emerging field has been discussed as "Neo-Craft" (Alfoldy 2010), "digital craft" (Oxman 2007), "hybrid" (Zoran and Buechley 2010), or "future craft" (Bonanni, Parkes, and Ishii 2008). All of them try to capture the relationship of craft and digital technology. What are we talking about when we speak of these hybrid craft forms?

The following sections outline a view of craft that connects to these debates. Craft is not applied as an ideology or a solution but as a rich reference of human practices for creative encounters with materials. The performance section of this book connected human expression to the collaborating materials in acts marked as performances. The craft section explores this connection in the generation of an artifact. This act is marked as making. But first, some definitions are required to clarify how a challenging term like *craft* will be applied.

Lucie-Smith (1981, 11) adapted his definition of traditional craft as a "calling requiring special skill and knowledge" and focuses on the role of craft "as a practical expression of social life" (7). Crafters are presented as individuals with special skills, who—through the performance of those skills—actively create and maintain social structures.

More recently, the director of the UK Crafts Council, Rosy Greenless, provided a layered definition: "Contemporary craft is about making things.

It is an intellectual and physical activity where the maker explores the infinite possibilities of materials and processes to produce unique objects. To see craft is to enter a world of wonderful things which can be challenging, beautiful, sometimes useful, tactile, extraordinary; and to understand and enjoy the energy and care which has gone into their making" (Westecott 2013). In her case, craft consists of intellectual as well as physical activities that do not end with the production. They extend into perception, use, and meaning making.

Taking an art historical approach, Adamson (2007, 2) looks at craft as a specific practice that presents a "conceptual limit" for art. Its practices can inform a creative approach to the making process itself. Adamson calls this "thinking through craft." He turns toward craft as a means in art, not to return into a "crabby conservatism" (2007, 168) but to pose a "problem to be thought through again and again" (168). Craft practices, changing relationships to materials, social context, and aesthetics are invited as critical lenses onto this production. This view is less concerned about the results, the final object, and instead centers on craft as critical practice. Adamson integrates craft into an art historical discourse and connects it to, among others, "relational aesthetics" (Bourriaud 2002) and "participatory art" (Kester 2004). As an end of means, craft is a productive practice leading to goods of certain quality. As an exploration of one's own obsessions and ideas, it is a cognitive-material exploration process.

Finally, craft is political. Morris's craft revival in the nineteenth century in England included a distinct turn to socialism. Gandhi's khadi movement in the early twentieth century used it as a form of protest for independence against the British Empire. Judy Chicago's *The Dinner Party* (1974–1979) works through craft as a feminist artwork honoring women's accomplishments. The story quilts of Faith Ringgold tell the unheard narratives of Black American women. The AIDS quilt, which started at the height of the AIDS pandemic, invited the public to collect memories of those who had died of HIV and to stitch and weave these stories into quilt patches that ultimately add up to an enormous collection that has been displayed on the Mall in Washington, DC. These are only selected examples of the way craft as creative practice can offer a voice today. Betsy Greer's (2008) "craftivism" or Carlton and Cooper's (2008) "Craftifesto" are only two modern-day instances of a movement that turns to craft as a means to express and demand political and social change. This spirit is reflected in the turn to a "handmade nation" (Levine and Heimerl 2008) or toward "handmade futures" (Press 2007) that envision different futures we could construct. This aspect of craft returns in the final section of this book, and though this section will center on the personal encounter first, it should be clear that any material encounter discussed always also has such a political dimension.

This brief review of craft definitions leads to a set of first keywords and qualities. Craft is based on material practices that depend on people's skills and physical engagement as well as materials, things, and context. Craft as practice expands beyond functional need and production to the skills and knowledge of crafters, who "think through" and express themselves. In doing so, they are coconstructing themselves through that practice, and they are participating in the shaping of the material as an active part in this process. Combining these criteria, our working definition of craft is *cocreative material practices* that are *based on needs and further individuation.* Within vital media, craft practices are *necessary, emergent,* and *constitutive.* They are based on needs, they unfold in the intra-active encounter of cocreative agents, and they further the individuation of all partners involved, human as much as nonhuman.

The following sections expand on these themes and first looks at the notion of need. From there, the argument turns to the role of craft for objects and individuation through practice. It relates to key elements of the craft definition and gradually draws it in relation to the digital. The overall goal of these sections is to explore craft as a practice to inform vital media production that emerge from material practices and eventually reach digital and hybrid designs.

Design and Craft and Need

The origins of practices in craft are specific needs. These include the need for a vessel to carry water, the need for clothing to withstand the cold, the need for food and its preparation to feed. Risatti (2007, 20) sees such a practical physical function as the "normative ground upon which craft originated." For him, this function serves as the unifying element that binds wildly diverging craft practices. Crafted objects, no matter whether they are made of wood, clay, or metal, share a functional quality that answers a particular necessity. Risatti lays out three main functions that craft has to answer to: containing, covering, supporting (32). They are part of a taxonomy of craft objects within which craft is clearly framed, and a craft object has to be "self-sufficient" to provide at least one of these functions. That means a craft object cannot require any external contribution to perform its function. For Risatti, a cup is self-sufficient because it can fulfill its function to contain liquid without any further functional aids. By the same logic, a piece of jewelry is not self-sufficient because jewelry adorns something else. Jewelry requires a body to be adorned in order to realize its operation and to achieve its purpose and impact (38).

Risatti's focus on need and on physical functionality recognizes the foundational value of an operational craft object. But it reaches too short to answer to the conditions of vital media. Differentiating between self-contained functionality and codependent functionality downplays the coexistence of any object in a shared and emergent world. In vital media, all functionality is part of a being together and of a collaborative coming into being. That means that self-sufficiency is an illusion. Dependency on another body is not a qualitative difference of jewelry compared to that of a cup or bucket. A cup depends on a plain surface to stand on and a hand to hold it and the laws of gravity. A teapot—certainly a self-sustaining object in Risatti's reading—depends on heat, on burning materials, on liquid to fill it to function, not to extrapolate on a culture of tea production and consumption. Looking at a teapot from the perspective of vital media unfolds an ever-expanding network including its production, shape, coloring, rules of handling, and who might be allowed to operate it and who not. The list of dependencies is infinite. We do not have to dig deep to realize that no object could be labeled functionally self-sufficient. Simply put, self-containment is not available in a collaboratively emergent world.

Secondly, if we accept cognition as a seamless expansion of other worldly forces, then the support that jewelry might provide for the individuation of cognitive beings is just as necessary a component as containing liquids might be for a cup. If our cognitive activity is part of the unfolding evolutionary project, if the human project is an integral part of the evolutionary one, then it also falls into the same need category. Need has to be defined by both the physical needs of materials and bodies as well as the cognitive needs of cognizant beings. As Greenless mentioned in her definition: craft is about making as well as seeing and understanding. Adamson's "thinking through craft" pushes this connection even further.

This critique of Risatti's original dictum expands outward to the needs of humans. But utilitarian functionality to support basic requirements remains a central component. To immediately step into cognitive engagement carries the danger of tipping the scale too far toward the importance of craft as a social act. As Lucie-Smith and others have noted, craft is a social construction force. Yet this does not replace Risatti's functional needs. Craft produces both, and through this combination, it is bringing forth the material and object as much as the individual and their culture. As Sennett (2008, 144) notes: "Marking an object can be a political act, not in the programmatic sense, but in the more fundamental matter of establishing one's presence, objectively." A potter working with the clay to form a teapot not only participates in the forming of the material, but they will sign the bottom of the pot, marking the human as much as the material presence. Both have emerged in this process. A society that builds around such processes grows in diversity and richness through the social and material construction. Rejecting such shared emergence is a sign of conflict to say the least.

As Welsch would argue, shared construction expands the world that brought it forward. This leads to another addition to Risatti's original notion of need. One learns from using a carving tool, no matter whether that is for the construction of a hunter's spear or the shaping of a shaman's mask. The hands shaping the material are also directly informing and shaping the cognition of the human (Wilson 1999). This learning is also part of the emergence that brings a shared world forward. It relates to the optimization of the technical object suggested by Simondon as much as the development of the human cognition. We grow as human beings with our tools. The field of embodied cognition covers one part of this growth. Any improvement to a carving tool or carving practice developed during the creation of an amulet will also inform the creation of the spear needed to feed one's body. Such an improvement can be functional in the sense of Risatti. It can include new materials, new forms that lead to new operations. Or it can be read as a cognitive expansion that includes the modification of practices, optimization of the way we throw clay or chip flint for arrowheads. In other words, it is part of a necessary learning process. Fulfilling these needs means to sustain, survive, and—over time—to evolve.

Making jewelry drives evolution, which we might call the ultimate function. A wedding ring, a scepter, or a neolithic death mask might not fulfill bodily needs in the same way that a knife, a pot, or a coat do. These objects remain dependent on their wearers to coconstruct identity between individuals. Yet they also coconstruct means and meanings of making. Over time, these construction processes operate beyond the individual or tribal or societal lifetime. These objects and the connected practices serve functions beyond physical needs, supporting memory and social construction. They become part of a material culture.

Csikszentmihalyi and Rochberg-Halton (1981, 231) look at the meaning of objects to suggest what they call "instrumental materialism," which "involves the cultivation of objects as essential means for discovering and furthering goals, so that the objects themselves may be cultivated through transactions." This reaches beyond consumption. The cultivation of the object stands "within a context whose purpose is the fuller unfolding of human life" (231). Their focus is on domestic objects and how they participate in different forms of cultivation. Within this field, they trace functionalities as assisting cultural construction and as tools for becoming with one's surrounding universe. "The objects that people use, despite their incredible diversity and sometimes contradictory usage, appear to be signs on a blueprint that represents the relation of man to himself, to his fellows, and to the universe" (38). Building these relations is an example of cocreative material practice based on needs and individuation. It also sees the function of objects in a wider context, expanding Risatti's original framework to include this kind of relating.

Risatti uses need to differentiate craft from art, but a focus on functional materiality—no matter whether it is physical, social, or personal—also differentiates craft from design. For a crafter, the making practice is part of their relation making to that object. In contrast, a designer's work centers on an optimized plan, a design, for the object. Craft conveys the practice toward an object; design conveys the practice of object making. Craft operates and speaks through material practices. Design speaks through communication about those practices. These differences might be clear enough but often blur in everyday procedures. Because crafters often have to explain their practices to clients, apprentices, or coworkers, they also engage in design. Whenever they lay out plans for a custom-built piece for a client or explain an assistant how to help, they communicate about the piece at hand. But during the actual making process, the workmanship of crafters takes over from any planning. One cannot design one's way to an object; it requires to be made. In that making and the encounter with the material, the skillful mastery of the crafter comes to live in dialogue with the material and the environmental context at hand. As Pye (1968, 62; emphasis in the original) argues, "*Design begins to fail to control the appearance of the environment at just those ranges at which the environment most impinges on us.*" Where Risatti emphasizes physical functionality, Pye focuses on the diversity of formal elements. These elements have to be practiced by the body in a craft practice. A skilled crafter has the ability to produce higher diversity in their encounter with the material. This depends on the actual making processes and cannot be conducted differently. In contrast, design "can be conveyed in words and by drawing" (Pye 1968, 17). Not unlike Pye, Adamson (2007, 4) supports this critical role of embodied skill: "Skill is the most complete embodiment of craft as an active, relational concept rather than a fixed category." Importantly, skill is not a separate ability or condition, but it is relational. It is not owned by the human but the result of the encounter with materials. Skilled practice and production are relational activities typical for craft. They are at the core of Adamson's "thinking through craft," and although designers might apply them to their practice, their home is in craft.

The terms of skill as embodied encounter and need as material and cognitive function both apply to vital media. In vital media, need includes necessary individuation and self-realization as much as the balance of basic body and object functions. Ultimately, Risatti's notion of need helps to realize this balance of coexisting and interdependent necessities. But it requires an expansion to include individuation in the list of these necessities. Making self, establishing one's presence, is an effort of becoming with, an embrace of the situated nature of our shared existence. It is not a stepping out of the coexistence or a departure from basic necessities. It is a cognitive need for which we collaborate with materials and objects through shared practices. Craft's qualities apply where such identity is constructed through cocreative material practices. The willed object, constructed with skill and purpose, becomes an active element not only through physical function but also in its application toward identity construction.

Ehren Tool's cup is a vessel that can contain water to drink, and this function is part of its nature. This fundamental function, the need it caters to, is indeed part of Tool's reason for producing cups and not jewelry. But at the same time, each cup is also a result and part of Tool's personal story and experience. It is a constructive statement beyond being a vessel, and it reaches out to speak to wider audiences. From the throwing of each cup to its changes in the firing, customization, and marking, to its distribution and life in the homes of the new owner, the whole process constructs identity. Such construction continues beyond the making process. Tool gives his cups away, often to politicians, other veterans, and widows, as well as random visitors to his events. For each of them, the cup can take on its own meaning as the story continues with new partners. As the cups' stories continue in new contexts, needs interweave, expand, and assemble through these developing interdependencies.

In vital media, the needs of the human participant cannot be separated from the needs of the rest of the world because of these interdependencies. Our stories are part of the world's stories. The material needs and functions are part of our needs and operations. This connects the defining relevance of need to the interdependencies of personal and cultural development. It defines the role of craft as the embodied practice of collaboration, as emerging skill, that manifests these interdependencies between noncognizant and cognitive beings. Media are networks of activity, and vital describes the quality of balancing the needs of the partners involved. Craft is the practice of this balancing in material encounters, and it covers physical necessities as much as individual cognition. This should clarify the important role of craft as practice for vital media construction. It states that craft is not applied as a historical looking back at a possible ideal but as a material and cognitive development process that exemplifies coemergence. But what about the resulting crafted objects?

Craft and the Objects It Makes

Through the lens of art, craft is often seen as a means toward an end goal. The goal might be a particular piece, a building, a piece of clothing, a painting. Craft is the way to achieve it. Craft is "supplemental" (Adamson 2007) as it deals with the changes to a material toward a finished object. It takes a lot of material knowledge to form, polish, and coat steel to perfection. But a view focused on the object as art would instead concentrate on the result. In that case, Jeff Koons's *Balloon Dog*, a metal sculpture shaped like a dog figure constructed from twisted balloons, is an art object. It certainly utilizes craft's efforts for its means but does not give it credit in the results, which seem to transcend the crafted work. Craft as a supplement remains applied to a piece that ultimately detaches from it. The artwork uses craft techniques, but its main expression manifests as the applied making dissolves into the desired effect. In this view, we look at a painting and do not see the mastery of the craft at work in the mixing of the original paints. Instead, we see the color red as a lively expression. We hear the sound of a Stradivari violin and appreciate it because we are oblivious to the mastery of the varnish and the forming of the body created in Stradivari's workshop. If we would focus on the varnish, the color as crafted, the steel as polished, the targeted effect of the musical or visual expression might be endangered. Such a view looks "through" craft to recognize the art. Once we are stuck in this line of thought, craft is quickly framed as "applied art" or "decorative."

Craft, here, only comes to the forefront through a change of perception, through some intervention or exceptional performance visible in the object. The joints of a piece of furniture can be realized as wonderous works when optimized in their technicality—as one can see in work of legendary woodworkers, such as Sam Maloof. The regularity of the thinnest possible thrown cup can become so extraordinary that it invokes marvel—as one can see in ceramic masterworks from China. Mastery on the highest level demands attention as the process pushes itself into the foreground. It reemerges through exceptional results that cause wonder and invite a change of how one might see a chair or a cup.

Yet efficient technical realization is not enough to cover the importance of the process in relation to the object. Reducing craft to technical competence is still degrading it. Increasingly, such proficiency can be mimicked by machines that devalue such an applied-like role of craft. Thanks to optimized industrial production, we are used to uniform and seemingly flawless objects produced by constantly improving technical mass fabrication processes. The most consistent thin ceramics are not produced by humans but by machines. Impossibly small versions are found, for example, in electronics used for circuit boards. The processes of their manufacturing follow design optimization to guarantee consistently high quality as well as commercial viability. Once such a design optimizes processes and forms, it can lead to countless optimized copies, countless cups with a specific shape, all seemingly of equal quality. That is why the Industrial Revolution eventually eroded the economic basis of handmade objects. In terms of sheer object productivity, technical precision, and economic efficiency, the mechanical reproduction can be fine-tuned to high efficiency and craft as the means of production becomes replaceable. It is debatable whether the handmade might still outshine the industrial production at the highest level of mastery, but in the area of daily commercial production, the battle was won by the machines long ago.

A single standardized paper cup will be cheaper and more energy efficient in its production for immediate commercial interests than a handmade cup. It stands in no relation to the efforts necessary for the individual cups created by Ehren Tool. If it does, it contradicts them. Still, Tool's cups are rooted in the same basic utilitarian function as the paper cup. Both incarnations are made to hold liquid. But Tool's cups are not optimized toward a standardized performance or appearance in their making nor in their unified nature as objects. They work, but they work individually and through more than their ability to hold liquids.

When design lays out optimized plans for making objects, it limits their variations, the diversity that Pye (1968) champions in object production. In contrast, handcrafted objects may include "slight improvisations, divagations and irregularities so that we are continually presented with fresh and unexpected incidents of form" (Pye 1968, 63). Tracing those signs of improvisation, Pye looks for irregularities on a sliding scale from the far to the near experience of any object. Unique features are encountered when one holds a handcrafted cup or wears crafted clothing. Earlier, Foerster posed the "number of choices" as the ethical imperative of second-order cybernetics. Now we realize the way that craft supports such diversity and how it is part of this "ethical imperative." Diversity is the range of different sensorial points of contact, and a richness of choices speaks not only to the level of detail in the nonhuman object but also to the senses of the human, which foster the cognitive engagement with the object and the world. A standardized paper cup provides identity mainly through size, function, and possibly a printed brand logo. In comparison, the diversity of a handcrafted item remains rich because of its unique size, shape, and color. This richness carries on to even the closest encounter with its rim, foot, and texture.

An ultimate bonus of this range of diversity on all levels is a gain in specificity. Specificity is the precision one can develop in one's relation to an object. Can you identify this item among other items? On which level of detail can this be done? Can others understand the distinction? It is not the identification of a brand or a concept but of the individual material thing. It is also not the recognition of the design of a particular object group, such as a cell phone or an automobile, but the specific item within that group. As the unique thing becomes recognizable, our engagement with it becomes richer. In that way, the continuous encounter with the material object continues to enrich and feed the cocreative process so central for vital media practices. For Pye (1968, 72), this kind of diversity brings back "something which is akin to the natural environment we have abandoned." Vital media do not cast a way back to the "abandoned," but it is clear that such an enrichment of specificity strengthens material connections and diversity. It allows for a richer position of self and object in the shared world and widens the options for the material culture that both inhabit.

Henry Glassie (1999, 41) defined *material culture* as the "tangible yield of human conduct." In material culture, "artifacts recall the technology by which nature was made cultural, and they incarnate the creator's mind, holding in form and ornament the plans that preceded them and the decisions committed in their making" (1999, 42). If media are networks of actions that include human expression, then objects can easily be read as manifestations of such actions, and human traces can be found in the making of such objects. Following Glassie, each of these objects can be read as a record of bodily action. "It incorporates intention," but "their reality does not depend on words" (44). These objects become wordless texts, telling stories of their creators, of their times and cultures. Through these stories, the objects literally position humans in the world: "Artifacts set the mind in the body, the body in the world" (42). This is ultimately an anthropological approach. Talking about the Turkish weaver Aysel Öztürk, Glassie tackles this transfer from mind to material. "That instant in which she translates herself into wool, when thought becomes material, is central, fundamental, and it gathers a host of associations that fuse in the act of creation. The associations are social" (51). Like most other scholars, he realizes the value of material, the contribution of the physical circumstances in the process, but his focus is on the reconstruction of sociocultural human conditions through objects that embed these associations. Öztürk's woven carpets are readable to Glassie as objects of material culture speaking of her family, her tradition, the world these practices are situated in. The call for vital media asks us to also see them as an expansion of the wool's characteristics, the tool's development, the diversification of the material actors.

As part of our digital media program at Georgia Tech, we created a weekly lecture series, where faculty presents ideas and questions to students as well as their fellow faculty members. Over a number of years, I presented the idea that a cup should be read as media. My emphasis on age-old craft was not easy to sell in a department founded to push digital media design and criticism. But Glassie would have no problem with that claim. For him, such objects as cups are part of cultural texts that avoid any abstraction into generalizable terms. This makes them uniquely relevant. They do not contain stories via words written on paper. Instead, they are individually marked, personal, concrete manifestations of a specific human action. The inconsistencies of Ehren Tool's cups, the dents left by his fingers on the inside, the form of the unfinished foot—these are traces of such manifestations.

Such a cup is not a replaceable operational commodity but a unique operator through the marks that inform about its creation. It is not abstracted, represented, and describable in words but precise, materialized, and captured in the specificity of the object. In contrast, commodities "are highly reduced entities and abstract in the sense that within the overall framework of technology they are free of local and historical ties. Thus they are sharply defined and easily measured" (Borgmann 1984). Borgmann (1984) argues that we are moving away from "things" and toward "devices" that serve to make a single commodity highly available while concealing the characteristic way it is procured. A machine-mass-produced cup provides for the targeted use in an optimized way and hides the production itself. Unlike Tool's cups, it tries to detach this function from social and bodily engagement of the thing except when it supports the abstracted notion of brand, leaving only the commodity in evidence. This might very well be the intention of the design, which might want to keep the attention on the logo of the company and not dilute that brand identity with any experience of individual items. But the results still shape our social fabric. The single-use cups standardized in size and design by global coffee shops are part of a cultural moment and of our inching closer to commodification. They tell abstracted stories of the brand, minimize individuation of the encounter, and maximize efficiency in the marketplace. They also are part of our growing environmental challenges and reflections of how we deal with them in our day-to-day life.

In comparison, a Tool cup is also a device for drinking, and its nature depends on this functionality. But beyond this operation it is also a social sculpture that is part of a material and cultural dialogue, which uses and activates individual and unique characteristics in the human as well as the cup. This marks the difference between vital media objects and commodifying media: vital media aim to support our connections to the world around us. These connections are strengthened through diversity, the specificity of an object, and the richness of our relations to it. Mass-produced media objects too often aim to turn materials into standardized commodities, which ultimately weakens any sustainable material culture. This does not mean that all media have to be handcrafted. There are other ways, including personal customization or biological diversity, that can provide for the necessary richness. What it does emphasize, however, is a turn to the specificity of each object and diversity in material. Near the end of this chapter, Amit Zoran's work on hybrid reassemblages provides one possible example for this.

We have moved from craft as a means, becoming invisible through gifting its power to an art object, to craft as foregrounded signs of mastery threatened by industrial fabrication and design, to crafted inconsistencies as manifestations of culture making in a shared and cocreative milieu. Along the way, we have stepped further and further into the idea of the crafted object as one that drives its own mediation activities. From those, diversity and specificity stand out as key qualities of vital media objects to distinguish them from commodified media approaches. Following the action-based view proposed here, we shift the focus to these processes in more detail.

Becoming Together

One defining criterion for folk potters used to be that they would dig up and prepare their own clay. Consequently, potters settled at locations that provided these resources. These crafters had to haul clay out of the ground, clean it, prepare it, and throw it on a kick wheel to fire it in a self-built wood-fired kiln using local glaze mixtures, before they could sell it in a shop, usually adjacent to the workshop, often to locals who depended on the wares themselves. To some degree, these practices remain alive today, and we will return to them. As has been noted above, they cannot compete with the modern production and distribution of ceramics as manufacturing alternatives. Yet, the practices themselves remain valuable because they embody the process of object construction as a way of identity and world building. These potters encounter the material on the land they farm and live on, explore these material's properties in relation to the given needs, and create the necessary objects in a close relationship with their immediate and experienced environment. As optimized production processes, these practices have been superseded. In terms of engaging with the surrounding activities to balance materials, environmental factors, and personal actions, they remain important. The following paragraphs look into these kinds of relationships and how they manifest in material making.

This look at material practices is not a revivalist turn to a somewhat holistic "good life" of former days. Operating a wood kiln requires a lot of firing material. Mixing glazes includes the handling of aggressive chemicals that can literally burn the potter (Burrison [1983] 2008, 90). Lead might have interesting effects as a material in the glaze's coloring but potentially horrible impacts on a human using the cup. We cannot optimize one component while disregarding the others. Encounters, explorations, and collaborations are dangerous. There are good reasons why craft traditions had to improve over time, and these improvements have not yet come to a halt. They continuously revise the way practitioner and thing relate. Practice is the way of relating to each other. The discussion of craft objects earlier identified key qualities on the nonhuman side of this relationship. This section explores some of the relevant qualities on the side of the human contributor.

Mary Caroline Richards (1966, 34) sees the personal becoming of the crafter intrinsically tied to that of the handling of clay in her hands: "Each moment bears life forward. It is as if the form that grows within our acts sheds each successive moment like a skin; it is as if the inner form which grows as a being within us is brought to maturity through the successive deaths of its material stages. It seems that the potter and his craft have had a special aura from the earliest times. Pottery is the ancient ur-craft, earth-derived, center-oriented, container for nourishment, water carrier. Experiences of centering and of personal metamorphosis grow within the craft." From the specific handling of the clay, Richards draws parallels to human development in the material evolution: "Personal transformation, or the art of becoming a human being, has a very special counterpart in the potter's craft" (36).

This process requires time for material to change, form to cast, as well as reflection and experience on the human side. Crafting an object is an embodied process over time—multiple timelines, in fact. The material story of clay includes tectonic shifts, sediments forming, pressure, water, and minerals assembling in particular structures. Once dug out, the clay has to be cleaned and prepared before it gets formed, dries and is fired, and different firing conditions can use variable cooldown times to create effects, such as seen in raku firing. Our encounters with clay are not abstract but tangible and concrete within a context that spans millennia and reaches into the far future. The clay needs time—so does the potter. As a project over time, it allows humans to participate in gradual steps from encounter, to exploration, to cocreation during this collaboration. These time requirements clash with the speed of modern media, which often minimize time efforts through optimized material production. On the one hand, prototyping techniques push craft practices further, as new tools allow craftspeople to work *"at a speed that keeps pace with thinking and development of ideas"* (Brian Adams quoted from Bunnell 2004, 13; emphasis in the original). But if we discuss making as a critical process, then removing the slowness of the hand from the reflection process does not just speed it up. It *detaches* it from the material component. This detachment can have serious consequences.

Traditional, slower handicraft techniques often connect with the search for a communal, at times decisively noncommercial, maker alternative. Firing a kiln can be a social happening, an actual itself. In contrast, ever faster media production is a trademark of the accelerating narrative of the web and digital media development (Anderson 2012). It has been criticized repeatedly. Paul Virilio (1995) detected a "loss of orientation" and of local history in the new time conditions of the digital space. Even critics of a revivalist's idealization of craft, like Adamson (2013, 165), acknowledge that "craft stands in tacit opposition to the rapid-fire movement of contemporary life, especially life online." Indeed, the craft practices Richards talks about allow for the necessary time to reflect on any step of the process. They provide the structural means for the encounter, exploration, and cocreation. Each one requires time to be mindful, to listen to the material, but also to give time for the material to form and unfold its agencies. The credo to "follow the material" requires adjusting to the material's temporal frame. Following cannot be reduced to the human activity or experience, but the exploration is also one of a shared temporal practice. Even though there are no indications for a slowdown as the digital production phases into personal fabrication, vital media ask us to accept a slower and wider time frame to support a shared temporality. This is a design challenge that is required by the material object side as much as the cognitive human side. Vital media design needs to include the time to reflect and counter a dematerializing and dehumanizing temporal acceleration.

Computer-assisted manufacturing is praised as a powerful creative tool for prototyping that does not depend anymore on repetition and grinding manufacture. The ideal fabricator is praised for its ability "to make anything, including itself, by assembling atoms. It will be a self-reproducing machine" (Gershenfeld 2005, 4). This power is seen supporting new levels of creative making, where innovation trumps conservation. Expressing one's creativity through novel objects produced with amazing speed is held up as proof for the power of fabrication labs and maker spaces. Craft stands in contrast to that. It requires grinding repetition of the same patterns or activities. Forms and patterns are not supposed to change in each instance but are followed through over long periods, sometimes over generations of crafters. As Adamson (2013, 142; emphasis in the original) argues: "In its purest technical form, it is the opposite of creativity. It is premised upon the *absence* of originality." Originality, in Gershenfeld's view, celebrates the prototype. Even though crafters work with prototypes, a telling sign of successful craft is consistency in the quality of basic shapes over long periods of production. Such consistency depends on optimized practices.

A production potter throws clay by the pound. Originality in the form is reduced, and creative variation is limited for the sake of optimized production for a given need. A community has a certain storage capacity to make ends meet in time for the harvest and to be ready for the winter. Delivering on that need was the task, not originality of the individual piece. The output can still bear the potter's signature and establish the individual potter's presence, each piece still carries the marks of the hand, and each piece is a sign of the working with the material. But the uniqueness of the resulting object is an unavoidable side product of a practice that centers on a cocreative material encounter. The results are not novel interpretations of what a jug might look like but a great many jugs that vary in material realization and differ only minimally in their design.

Sennett (2008) introduces the term "material consciousness," which includes the three key issues of *metamorphosis*, *presence*, and *anthropomorphosis*. For Sennett (2008, 120), metamorphosis includes changes in the making (e.g., new production methods); presence includes the marking of the object by the maker (e.g., a brickmaker's stamp); anthropomorphosis "occurs when we impute human qualities to a raw material," when we endow inanimate things with human qualities. Defining practices this way, constituting the presence of the individual, is most fitting for a discussion of the craftsperson, which is Sennett's main concern. But it remains centered on the person. In the case of vital media, activity, individuation, and emerging context are not limited to nor centered on the contributions of the human. Projecting anthropomorphic qualities onto a material can even endanger that material's own agency. If we make sense of a material's action solely through the lens of human qualities, then we are bound to miss those qualities and actions that are not within our human realms but nevertheless central to the material. If we read the creating powers as belonging to the human only, then we are neglecting the material agency of others.

This shift away from the human center collides with varying interpretations of another key concept in many discussions of practice: creativity (for a discussion of definitions of the term *creativity*, see, e.g., Sawyer 2012). Creativity in craft is not about the realization of the "genius" of the human contributor; nor is it purely about the value of one's creative act for personal development or for sociohistoric advancement. All of the above might be affected, but creative action in a vital media production has to incorporate the roles of all actors. This includes tools, materials, workshops, and communities as well as human makers. Creative action in vital media is the realization of all components' evolution. They intertwine, as scholars like Sennett have powerfully argued, but they remain their own and need to be recognized to be balanced. The humans' individuality (presence) is an important part yet still only one among countless others. Projecting human qualities into the materials (anthropomorphosis) is a cultural construction that might very well emerge but is not part of the immediate cocreative encounter with the material, and we need to tread carefully to avoid material neglect. The development of new practices (metamorphosis) covers the emerging procedures but should not be understood as performed by humans in control over obedient materials. It always is a collaborative construction of both.

These encounters are transformative for all participants as they coconstruct each other. They are closer to Richards's understanding of metamorphosis. These metamorphoses are individual forms of self-construction for participating humans in their encounters with material, and, at the same time, they are a material's adjustments in its encounter with the human. Human and nonhuman assist each other in their codependent emergence. In this *"conservation of being through becoming"* (Simondon 1992, 301) we see traces of Simondon's "individuation" at work. Individuation sees any development as mutually transformative, as evolution in between partners, as dialogue. It can happen between crafters and materials as well as between crystals as they form. "Such an individuation is not to be thought of as the meeting of a previous form and matter existing as already constituted and separate terms, but a resolution taking place in the heart of a metastable system rich in potentials" (1992, 304). These potentials are provided in energy, form, and matter that allow the actualization of the crystal object or the clay pot. Like the metastable system that allows crystals to grow, the craft process (and we can add here: the vital media process) is a balanced exploration of shared potentials. "The new magic will not be found in a direct expression of the individual power to act, assured by the knowledge that gives each gesture effective certainty, but in the rationalization of forces that situate man by giving him meaning within a human and natural ensemble" (Simondon [1958] 2017, 119). Finding meaning in this ensemble has been the aspiration of many models. Vital media do not directly target any particular new meaning, but they support the "forces that situate man" and nonhumans in relation. They focus on the intradependencies as the shared activities and playgrounds of emergence.

We have listed time and mutual interdependencies as important conditions to arrive at coemergence. Furthermore, we have turned from a focus on practice as something performed and experienced by the crafter to activity that is always only relational, based on varied potentials, which cannot be described only by human attributes. Vital media do not see these potentials as limited to whatever might be preexistent in the system and instead see any system as open. The call for a metastable view of individuation of material and self still applies but there is no resolution to it. This opens up the question of defining the result of such practice. If we are centering all activity in an ever-emerging sea of agency, how do we identify some outcome? And how can such an outcome relate to the expectations?

In the craft-driven media production laid out here, the concept of a conclusion or end state is indeed disappearing. Craft-based media production is the never-ending individuation of cognitive and noncognizant beings that depends on all partners involved. It brings forth the potter and the cup through a material dialogue called practice. By definition, this process remains open. We cannot presume conclusions in an ongoing cycle of individuation. This allows a dialectic opening. Facing multiplicity and interdependencies in such open cycles is at the core of critical practice. It is also at the core of vital media practice modeled after craft. Following Simondon, we can trace this process to a form of individuation where the material and technological receive as much formative say as other factors, including the crafter. This is not a step back to technological determinism but a realization of the multifaceted construction and mediation powers of craft. Simondon particularly emphasizes that such individuation cannot presume any conclusion or projected result. The material does not project what it will become; neither can humans presume that their development is conclusive; nor can we presume a specific cup from the encounter of a potter with clay. Instead, it remains a process-based development. According to Simondon (1992, 300; emphasis in the original) we have to *"understand the individual from the perspective of the process of individuation rather than the process of individuation by means of the individual."* We have to embrace the process first and cannot frame it with a possible outcome or fixed goal in mind.

Hannah Perner-Wilson is a professor of digital media at the Department for Spiel und Object (Play and object) at the Ernst Busch University for Performing Arts in Berlin. The school features programs on choreography, acting, dance, directing, and puppetry right next to its newest offering on digital media. The new play and object program was established by Friedrich Kirschner and aims to integrate novel technologies and their practices into performance. Its focus on objects includes digital as much as material things, and it applies forms of play to mediated participation. Perner-Wilson's own work often combines these strands in various ways. She is a maker of physical objects as much as of digital connections, and the way we work with those objects is often performative, playful, and exploratory. The seamless bridges between expression through performative objects to crafting and interaction design make her work so remarkable. At the same time, her practices can serve as one example for how the call for material-inclusiveness can be brought forward—in other words, how to achieve a collaborative balance through practice.

Perner-Wilson's work often involves fabrics or some form of wearables. For a while, she and Mika Satomi ran a commercial tailor shop for e-textiles, which provided handmade clothes that incorporated various electronic and digital features. This might involve gloves that detect hand gestures or garments equipped with light sources that activate in patterns triggered by special forms of embodied interaction design. Such specialized work requires unique bespoke couture, but more importantly, it also depends on material explorations. In many ways, the resulting—often extraordinary—pieces represent only a fraction of the actual work. Underneath it, Perner-Wilson tackles materials—their properties, tools, and creative practices. This

Figure 4.3
Perner-Wilson's *Ohm-Hook*: for the use of conductive material in crochet. Used by permission of Hannah Perner-Wilson.

foundational work includes crochet tools for conductive thread (figure 4.3), dyes that make fabric conductive (Honnet et al. 2020) (figure 4.4), wearable tool vests (figure 4.5), and countless instructional videos and documents that share the underlying designs and making processes.

Perner-Wilson's practice is open and explorative. It remains closely tied to the materials she works with and is surrounded by. It explores physical affordances of materials as much as potential digital connections and the shared environments they might unfold in. It grants the nonhuman participants space, time, and opportunity to show and unfold their agencies. At the same time, it supports our roles as creative makers through unique practices and encounters with novel artifacts. One of her designs is a vest optimized for mobile use and tailoring on location, which further evolved into a prototyping lab vest supporting in situ rapid prototyping in all kinds of different environments. Her studies often quite literally combine explorations of the environment, such as a tailor's workshop or a tropical rain

Figure 4.4
PolySense: technique of polymerization to create and apply a conductive dye. Used by permission of Hannah Perner-Wilson.

forest, with that of novel materials, tools, practices, and our relationship to them. Her work is relational at its core and she accepts the importance of materials is tools in return: "My relationship with the textiles and electronics that I've spent the last ten plus years of my life with have gotten me wondering about the extent to which my world is shaped by the materials and tools that I work with" (Cole and Perner-Wilson 2019, 109–110).

The "Performance Makers" section focused on the inclusion of material objects in the production of expressive action. It defined performance as the action to produce expression with an intent. The definition of craft as cocreative material practices that are based on needs and further individuation still includes intent and production but has turned the perspective around to look at a physical making process first. In both cases, the collaboration has become the center of attention. Because the partners involved here feature their own time conditions and their own interdependent agencies, this center of cocreation loses many of its traditionally human-scoped qualities. It does not have to be fast to be inventive, it does not have to be conclusive to be productive, but it does have to contribute to the individuation of all partners involved. The resulting objects might never resolve the

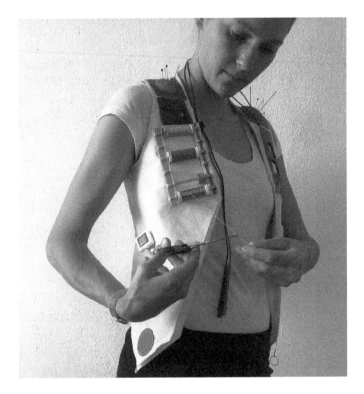

Figure 4.5
Wearable Studio Practice: customized vest for on location tailoring. Used by permission of Hannah Perner-Wilson.

challenge at hand, but they are always important results in and of themselves. This describes the concept of making as a *becoming together* that is multilayered, emergent, and interdependent.

Next, we turn to the forms of engagement that such a becoming together requires from the human participant. How does one act in these unresolvable conditions? And how do we make sure that this making activity is not only shared but also reflective in order to support critical development? The opportunities for a critical dialogue are set, but how do we realize them through our practical encounters?

Complicating Improvisation

Craftspeople plan to structure their work. They optimize their workload as well as their resources. The idea of reflection is not new in these planning processes. John Dewey (1910) championed "reflective thought" as a way to learn from experience, developing it into "reflective activity." With Dewey, the term got adopted and varied within education (Boud, Keogh, and Walker 1985), and this educational turn proposes that the material practice of making includes the reflection and education of the maker as much as the shaping of the material. Crafters learn as they work and modify their practices hands on. The materials in their hands cocreate these practices and change their conditions. Learning by doing is a give-and-take.

This also applies to craft. "Trying to separate cause from effect inside the loop of pottery making is like trying to construct a pot keeping your hands clean from the mud," argues Lambros Malafouris (2008, 25). For him, agency and intention are properties not of the partners but of the unfolding action itself. It is this "dance" that we need to explore to find our own footing. To do so, we have to interrupt our own stasis, break the security of an established routine, and destabilize ourselves in relation to the material. The process itself remains intrinsically interdependent as the agency in this interplay of making shifts and varies. This leads Malafouris (2008, 19) to argue that "underlying the effortless manner in which the potter's hand reaches for and gradually shapes the wet clay lies a whole set of conceptual challenges to some of our most deeply entrenched assumptions about what it means to be a human agent." Ingold (2013, 21) reduces the role of the human even further: "The most he can do is to intervene in worldly processes that are already going on, and which give rise to the forms of the living world that we see all around us . . . adding his own impetus to the forces and energies in play." This section looks into this transformative moment and how it can remain reflective and critical to explore our role as participating humans in it. It will break this moment into the three-step logic of *encounter, exploration,* and *collaboration.*

To describe a crafter's process, Charles and Janet Dixon Keller propose an activity system following Alexei Nikolaevich Leont'ev, which includes the formation of an "umbrella plan" and indicates a continuous feedback loop. "Action continually brings the human into practical contact with objects that deflect, change, and enrich prior organizations of knowledge" (Keller and Keller 1994, 126). This kind of "practical contact" is a constant reencountering and related to Dewey's "reflective activity." The skilled crafter invests their expertise but depends at the same time on the object's activities. A helpful focus in the wide range of emerging action is on the moment where the planned encounter is destabilized, when there are glitches in the system or encounters with unexpected materials. These are the moments that require improvisation to continue the cocreative material practices of people and materials. Dealing with such irregularities, glitches, "deflections," and failures is a defining element of Pye's (1968, 20) definition of craft: "It means simply workmanship using any kind of technique or apparatus, in which the quality of the result is not predetermined, but depends on the judgement, dexterity and care which the maker exercises as he works. The essential idea is that the quality of the result is continually at risk during the process of making; and so I shall call this kind of workmanship 'The workmanship of risk.'"

The Kellers' continual encounter with the undetermined relates to a key term of the material encounter: *surprise*. Surprise is born from risk and indicates necessary cognitive adjustments. Materials "deflect, change, and enrich," and these are moments of possible failure in which the cocreative forces are especially visible. These moments materialize when a potter feels an inconsistency in the clay, when a woodworker has to deal with irregular grain, or—as in the Kellers' case—when a smith has to adjust his work on the fly while the iron is hot and malleable. Ingold focuses entirely on this moment as he (2009) proposes to look at "real" objects as constantly coming into being through relational networks. Yet the first step for the human collaborator is to dissolve oneself into such networks, to let go of "safety," to embrace the "surprise" by accepting the "risk," and to subsume any planning to this new "enriching" entanglement. This describes the encounter between human and material, which not only is a physical meeting point but also destabilizes its own securities and preconceptions.

The next step is mutual *exploration*. This is the region covered by Donald Schön's (1987) "reflection-in-action," which "is a reflective conversation with the materials of a situation—'conversation,' now, in a metaphorical sense. Each person carries out his own evolving role in the collective performance, 'listens' to the surprises—or, as I shall say, 'back talk'" (Schön 1987, 31). Expanding Schön's point, the "back talk" goes both ways: material speaks as much as it listens in this "conversation." The operators are not finished constructs that speak from fixed points of completion but dynamic forces constantly at work. Ingold and Hallam (2007) describe such creative practices with material as *generative*, *relational*, *temporal*, and ultimately *improvisational*. Improvisation is a most fitting term to describe the human side of exploration. In it, practice is a collaborative process of bringing into being that is shaped by in-the-moment encounters with others and with material. These encounters explore not only the material but also the crafter's role and identity. Much like an improvisational actor constructs a role and identity on the fly as they enact it on stage, crafters construct their maker identity through their creative practice in the workshop.

Pye's workmanship of risk served as a principle to outline the encounter, and the notion of risk in improvisation remains a key element. The very real danger of falling flat, to "die" on the stage, is part of a good improv act. But in the mutual exploration, the vital media reading of craft differs from Pye's risk of failure in two points. First, craft is not only about the construction of the object but also of the crafter—thus risk needs to also include the failure of this personal individuation. Not unlike the risk a stand-up performer takes when entering the stage, the crafter encounters possibilities and failures when approaching the material. The crafter's needs, their cognitive development, might not happen. The crafter is as much at stake as the material.

Not unlike Fischer-Lichte's autopoietic feedback loop, this approach is a bodily—relational—encounter that feeds on the emerging forces between all partners. But it can fail on the side of the crafter as much as on the side of the material. It puts the crafter at risk as much as it does put the material in danger. Most of today's media practices do not support this approach. The famous Bauhaus artist Annie Albers had to escape the Nazi regime to eventually join the Black Mountain College in North Carolina. There, at the end of World War II, she wrote about the need to turn to a hands-on educational approach to learn how to make a stand, how to develop courage and independency:

> We come to know in artwork that we do not clearly know where we will arrive in our work, although we set the compass, our vision; that we are led, in going along, by material and work process. We have plans and blueprints, a shorthand of material and its treatment, but the finished work is still a surprise. We learn to listen to voices: to the yes or no of our material, our tools, our time. We come to know that only when we feel guided by them our work takes on form and meaning, that we are misled when we follow only our will. (Albers cited from Halper and Douglas 2009, 8)

If absolutely no personal development is happening, no reflection, no enrichment and growth, no development of tacit or technical knowledge on the crafter's side, no improvisational construction of one's own role, then the process is incomplete as craft even if the object is perfect. Then, the process has failed the crafter.

The second difference to Pye is to acknowledge that the controlling forces are not only the crafter's but also the materials'. Risk is not a failure of a crafter's skill applied to a material but a failure of the collaboration of the two. That means that a failure of the crafter's skill might still lead to a productive process with the help of the materials. Sometimes materials take over the relational banter during the producing discourse. In these moments, the materials take the wheel with surprising results. This reframes what might be seen as success or failure. Typical examples of this effect are lucky accidents that reveal new opportunities despite the crafter's "failure" to perform. The power in these relations do not simply reside only in the hands of a crafter. If the encounter destabilized maker and materials to open them up for an intradependent engagement, then this exploration is the manifestation of the shared and mutual making process.

If we recognize these dynamics and interdependencies, then an artifact is not a settled proof or final state. Instead, it is constantly changing and adjusting, and it is part of a larger dialogue far beyond a single manipulation or result. Ehren Tool's cups continue their conversation, they are becoming even years after they left his studio. Given the lifespan of fired ceramics, a great many of them will keep on "talking back" for multiple human generations to come. Exploration is a becoming in risky conditions for both, the material as much as the crafter. If the encounter stage was destabilizing the approach of each of them to the other, the exploration is assembling both on the stage, the workbench, the pottery wheel and defines them in their activities onto each other. After encounter and exploration, this leads to a third stage: *collaboration*.

Collaboration is the shared production toward something new. It is the basis for a new differentiation and individuation. In this way, collaboration leans on Barad's work on the "agential cut," which happens as active partners differentiate from each other as they intra-act. The encounter of different actors is an ongoing exploration for each of them. They literally bump into each other, affect each other; they enact materialization. They differentiate by intra-acting. Haraway (1988) describes such encounters as embodied mapping exercises in which new boundaries are drawn, new objects emerge, and new identities constructed. Extending these further into active materials, Barad (2007) uses the example of the brittle star to emphasize that "differential materialization is discursive." The ongoing bodily reconfigurations of the brittle star "are products of iterative causal intra-actions—material-discursive practices—through which the agential cut between 'self' and 'other' (e.g., 'surrounding environment') is differentially enacted" (Barad 2007, 376). We cannot speak of the producer of an object or an author of a text or a user of an application or a player of a video game. We can only speak of the discursive encounter, and along its fold lines, the differences engage, and the new production of brittle stars or humans or video games happens.

In my classes, I often use assignments that present conceptual challenges like these to students, and I regularly marvel at their responses. Claire Stricklin found herself in such an assignment that asked her to build a digital media intervention based on our bodily encounter with a particular material of his choice. She found inspiration in Jim Melchert's piece *Changes* (1972). For *Changes*, Melchert invited friends into his ceramic studio and prepared a bucket of slip. Slip is a liquid mix of water and clay used in ceramics for many purposes, including the connection two pieces or to decorate clay shapes. Instead of using the slip as material on other ceramics, Melchert invited his visitors to dunk their heads into the slip and let the material gradually dry on their own bodies (see figure 4.6). In that way, participants experienced a change in perception over time as the slip closed their ears, changed its temperature while hardening, forced them to keep their eyes closed. While we are used to seeing the human affecting the clay in ceramic arts, *Changes* focused on the changes that the clay applied to the human. But Stricklin had to combine the material encounter with a digital component, too. She combined the inspiration from Melchert with her own experience as an avid player and author of tabletop role-playing games, such as *Dungeons and Dragons* (Gygax 1974).

Ultimately, Stricklin decided to build her project around Hydrostone and its material properties. Hydrostone is a gypsum cement that air dries into a very hard structure. It is used to make details or casts of model components. Tabletop role players use it to build parts of their fictional worlds. They cast components, such as walls, arches, or other details of the maps they use in their play campaigns. But Stricklin not only plays; she also writes campaigns for such adventures and helps to invent the immersive worlds for players. Her project, *Primal Clay*, came together as its own world-building game. As the main digital component, it used an accumulative interactive fiction.

Figure 4.6
Jim Melchert during *Changes* (1972). Photo by Mieke Hille; used by permission of Jim Melchert.

This largely text-based component guided players to imagine and perform loosely suggested actions. It provided prompts and gradually assembled the resulting text summary from the player's choices. On the physical/material side, these prompts asked players to act in collaboration with a batch of Hydrostone and perform a literal world-creation exercise. Stricklin's game starts with Hydrostone being cast on a play surface generating a kind of sandbox to form a world in, as seen in figure 4.7. Players are prompted by the interactive fiction to enact changes to the Hydrostone, to impose world building events such as cataclysmic disasters or developmental changes.

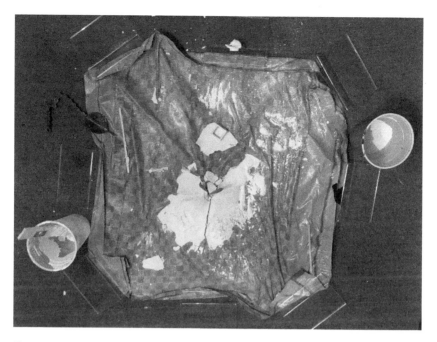

Figure 4.7
Primal Clay: play surface. Photo by Claire Stricklin. Used by permission of Stricklin.

All the while, the material itself rapidly dries and hardens. The gradually hardening Hydrostone effectively works like a timer controlling the different stages of the play in combination with the digital fiction fragments. Its changing affordance—first a rather liquid slip-like material that eventually solidifies into a hardened cement—affects player actions throughout in a dynamic way. Any enactment of the digital prompts is already a messy enacted improvisation with the Hydrostone, but the effects become even more obvious in random encounters. One "meteor" prompt calls the players to take turns and "drop a heavy object onto the world." During one play testing, the players used a flower arrangement as their "meteor." It did not do much damage to the nearly hardened Hydrostone world anymore. But it dropped flower petals, which immediately were integrated in the world-creation lore as enormous flowering vines rising high above the world seen in figure 4.8. *Primal Clay* is a speculative game prototype. It lacks the balance that makes *Dungeons and Dragons* such an alluring system, and its design is more reaching than definitive. But as a material-hybrid design exercise, it manages to highlight valuable key points. The collaboration

Figure 4.8
Primal Clay: flower petal forest. Photo by Claire Stricklin. Used by permission of Stricklin.

toward the creation of something new—in Stricklin's case a new "world"—emerges from the forces of all active participants as they meet. Digital prompt meets players' physical action; vase meets Hydrostone; flower petal meets fictional world. Each emphasizes the contributions of the individual but toward the formation of something new.

The Hydrostone worlds do not really end, either as the *Primal Clay* fiction calls players to include past components into the emerging new worlds. It encourages players to connect constructed world upon constructed world. This might be informed by the reuse of comparable materials in a *Dungeons and Dragons* campaign. The models of a fortification in a model castle might be reused as parts of ancient ruins for a different map environment the next day. This is the nature of a toolbox, where one cannot presume a single final object from such a playing and crafting process. *Primal Clay* is as much a system-based game idea as the tabletop role-playing games it derived from, but it manages to integrate the voice of the Hydrostone into these systems. Vital media have outlined this encounter as a dialogue that is transformative as it fosters the individuation of all partners involved. We step forward together as we form individuals, objects, "phenomena" as Barad would claim, between us. This is the moment of a successful collaboration, in which intradependency supports construction of object, identity, world, or idea.

Particular qualities of craft stand out in this shared construction. This section discussed a range of them: different speeds and repetitive practices that allow for reflection, material feedback loops that support emergent transformation and learning, recognition of the interdependent emergence of maker and materials without closure for either, improvisational responses feeding these relational practices and opening up to further reflection. It argued that vital media are expressions to construct individual identity as much as material transformation shaping stuff. They require the concretization of the object as well as the individuation of the human participants.

These relational practices can be structured into *encounter*, *exploration*, and *collaboration*, and this chapter has laid out their relationship with an ongoing shared practice. Whether it is through planning or in improvised responses to moments of surprise, for this collaboration to be successful and sustainable, it needs to be critical. This gives rise to the next question: How does making become critical?

Critical Partners

In *The Craftsman*, Sennet (2008) suggests three core abilities for craftsmanship: *localization*, *questioning*, and *opening up* to discuss constructive encounters with materials (Sennett 2008, 277). *Localization* is a form of sensory input: assessing the smoothness of a wooden surface via touch, feeling the wetness of the clay between one's fingers, attending to the proportions and shape of a piece of jewelry. Focal attention is brought to a material, process, or object to assess details to deal with. *Questioning* investigates opportunities at hand and matches them to existing practices and knowledge. Sennett (2008, 279) likens this reflecting on a material's qualities to a cognitive state of "dwelling in an incipient state" in which "the pondering brain is considering its circuit options." It also relates to "operational success" (279) when logical connections are realized during tasks such as programming. *Opening up* uses domain shifts to further activate, even reframe the encounter. A change of habitual practice is one way to open up new actions toward the task at hand. This might include realizing that a tool made for hammering might be valuable as a lever, too. A framework developed for one domain might apply to unexpected new ones, as seen in the application of patterns developed by Christopher Alexander for spatial design onto computational logic. *Opening up* relates to "doing things differently" (279). These three steps relate to the aforementioned phases of encounter, exploration, and collaboration proposed for vital media practices. Sennett's work clearly recognizes the importance of materials, but he still focuses on the crafter. It is the human who performs localization, questioning, and opening up. For Sennett, craft assists the cognitive evolution of human development, and materials are crucially important for that. But they are a means in that advancement and remain underrepresented as actors in and of themselves when it comes to the critical encounter.

Likewise, for Adamson, the idea of thinking through craft includes the turn to higher cognitive functions. Critical theory and reflection become part of our encounter with the materials. Here, craft is a relational exercise in which humans meet various elements from materials to tools to objects in the pursuit to further their own thoughts. The encounter is focused on human logic, on thinking, memorialization, even trauma. At one moment, Adamson (2013, 186) points out that "the syntactical structures of craft and storytelling are closely parallel" as he develops the value of craft for memory work. But narratives require cognition, and a focus on storytelling is by design also a focus on the human participant's side.

A related view of craft as intellectual engagement leads Dormer (1997, 152) to see it as a "form of intellectual and imaginative possession." For him, "making is both the means through which the craftsperson explores their obsession or idea and an end in itself" (1997, 154). Craft is both personal and part of a nonhuman economy and material environment. "It is not craft as 'handcraft' that defines contemporary craftsmanship: it is craft as knowledge that empowers a maker to take charge of technology" (140).

All three approaches frame craft as a critical form of engagement driven by the crafter. They are potent references to interrogate our material practices and question media design based on material encounters on many levels. Does the person have the opportunity to reframe, to "think through" the form of interaction that a digital media device provides? Or is the interaction rigid, using the material encounter as a wrongly fixated one? Does the interface allow for a sensing, a localizing of the tactile encounter? Or is there an inherent mismatch between what the hand feels and what the operational function does? Does the interaction empower to take shared improvisational action? Or does it subsume to a prefabricated plan? All these are powerful considerations that have shaped such fields as tangible interaction design, game design, and human-computer interaction at large. Because craft allows for a personal reflection, for a development of ideas and a thinking through practice, it supports critical dialogue from the human participant's side.

This all still applies to vital media. Materials still allow us to question conditions. But vital media also demand that we widen such a view to look beyond the human. Craft as a reference to media design is relevant not only because it highlights the physical and cognitive activities of humans but also because it provides voice and presence for the agency of materials. It realizes the cocreative efforts of all partners. To repeat the credo of this book: we need to assist the construction of all the collaborating partners as they intra-act with each other.

The performance section started from a field that traditionally emphasizes the ephemeral human expression. In this case, we emphasized material as a cocreative force in the emergent media event that gradually includes material voices. This was discussed in a brief look at puppetry but also traced in digital formats, like the cell phone app for the *Subway* project. If the performance section targeted a correction of human expressions into material encounters, then the craft chapter hopes to deliver a counterbalance to correct a view of craft as the application of human will onto a material resource. It argues that crafted objects remain active parts of an expressive as well as productive emergent dialogue. Crafting itself is a form of individuation of both, the human partner as much as the materials and objects that emerge. The human experience in this process has been covered at some length in scholarly work, but the active material component has found less attention. That is why this chapter is titled "Recentering." Its narrative shifted the focus from the human's shaping hands toward the emergent cocreative practices *between* partners. If dependencies and practices are truly intertwined, then we cannot limit critical action to human reflection about the material engagement. Cognitive expansion and material diversification do not "come out" or detach from each other. Both constitute each other through their interoperations. They perform this collaborative constitution because of interconnecting needs and abilities. They support a critical encounter that is formational for all who are involved.

Vital media designs and their application of craft ideas are not aimed at the improvement of humanity but at the improvement of the conversation, the material-cognitive evolutionary project of a shared evolving existence through differentiation. The turn to craft provides a method to critically reflect, explore, and reframe while keeping all components actively engaged in the practice. Interacting with vital media is not about the mastery of a particular practice. It is about a critical encounter with specific materials as such. The past sections argued against the view of craft as an ideal mastery over a material and against measuring critical engagement by that standard. Vital media identifies willed expression as a cognitive human contribution, but through craft, we come to realize the coconstitutive interplay between operating materials and people. We also realize that no conclusive mastery is ever achieved and that raising the human skills over the material's contributions would reverse the balance we are trying to strike. If the critical operations in this balance are meant to stay active, then the argument cannot end on any conclusive skill set or preset critical framing in these interdependent construction cycles. We can conduct encounters, explorations, and collaborations on every skill level with always differing yet valid results. Thus, the argument turns to education.

The concept of *critical making* proposes to bridge the gap between our conceptual perception of technology and our material encounter with its objects "to use material forms of engagement with technologies to supplement and extend critical reflection and, in doing so, to reconnect our lived experiences with technologies to social and conceptual critique" (Ratto 2011, 253). For Ratto (2011, 259), who coined the term, "critical making is about turning the relationship between technology and society from a 'matter of fact' into a 'matter of concern.' I see this as requiring personal investment, a 'caring for' that is not typically part of either technical or social scholarly education." It is an educational project aimed at fostering such a form of caring, and it does so through "material engagement" and "negotiations" with objects. Different practitioners apply critical making in their own ways. Garnet Hertz, for example, breaks his approach down into five steps: (1) identify disciplinary metaphors and assumptions (what are key assumptions we have in a given discipline?); (2) research metaphoric occlusions (how do existing assumptions occlude groups?); (3) invert occlusions (bring the occluded to the center in your design); (4) build the inversion (create the physical object); (5) deploy the object (Ratto and Hertz

2019, 25). In practice, it often takes the form of workshops that emphasize making exercises and shared reflections among participants. Instead of dividing critical thinking from making practices, the making becomes an integral part of the critical reflection processes. This logic builds on earlier approaches, such as critical design, critical technical practice, and reflective design (Ratto and Hertz 2019).

Critical design poses provocative questions to everyday life that are not aimed at solving a specific problem through novel products but at supporting a new imagining. They respond to the usability-centered focus of earlier design approaches, which targeted optimization for use and functionality. To counter such engrained perspectives, "it is not enough to simply offer an alternative, new strategies need to be developed that are both critical and optimistic, that engage with and challenge industry's technological agenda" (Dunne and Raby 2001, 59). The goal is to advance the "lived experience," which focuses on the social, cultural, and psychological world of users and communities. Critical design questions market values, social norms, or technology fetishism or targets any other given norm. Because it does not promise solutions, it can be highly speculative and conceptual. What would our society look like if fossil fuels were depleted? What if the ocean level rises? For speculations like these, any material encounter and object can become an evocative means to activate critical thought and reflection. These objects can be highly inspirational as they form part of an argument that the designer opens toward the audience. The materials are part of building the argumentation, but the audience for that argument remains human. The main goal is societal change and the designed object is a means to support that.

Critical making shares these aspirations. As a method, critical making is not by definition activism, but as an educational take on material-technological encounters, it lays the groundwork for activist engagements. The role of material engagement, of making with operational materials and objects in your hand, is more central in critical making than in critical design. It is a foundational part of the learning process itself. In this regard, critical making is based on constructionist pedagogy (Papert and Harel 1991) and experiential learning (Dewey [1938] 1997). It does not teach specific technologies or material practices with a view to their optimization but as forms of exploration. Critical making is not craft by definition, but it features the exploration mechanisms outlined above, and it shares an open-endedness with the kind of ongoing craft encounter suggested in this book. One never masters a production practice through critical making as the encounters remain preliminary and exploratory. These material encounters can remain messy, unresolved, and mysterious. Our relationship to them can be immature, at times even naïve, as long as they support our reflection processes.

But critical making centers so much on these human reflection processes that the voices of the materials are muted even though the encounter with them is marked as pivotal in the practice. We learn and realize new connections in the making, but the material's role is downplayed in the resulting achievements. The collaboration part is weakened in critical making as the material encounter is not meant to lead to any development. Diversification of the material or an "agential cut" for an emergent object is not central to critical making.

A critical making workshop might form teams of participants to explore a particular challenge, such as complexities of network systems or privacy in big data. These participants might engage in a hands-on making exercise and build robots (Ratto 2011) or soft circuitry in e-textiles (Kafai and Peppler 2014). These building exercises can successfully stimulate further reflection among the workshop participants. Thus, interoperating with technology helps the human participants grapple with their understanding of the original challenge concerning networks, privacy, big data, or whatever the identified problem was. But once this process is achieved, the objects that were made as part of a hands-on exploration are reduced to mere steppingstones. The human workshop participants used their making as a stimulant to learn. The objects might be evocative to others as educational references or expressive as critical design artifacts illustrating a particular idea, but they do not have to be. They do not need to work beyond the original conversation, and their own emergence as objects of value and necessity is diminished. The need for any object built in a critical making workshop depends on its educational impact—at least by Ratto's definition. Hertz's interpretation of critical making leaves more room for the resulting object to unfold in a public release. Inherent material necessities might support this educational impact and help to engage the human workshop participants, but they are not important beyond such reflection and activation. Their voices are not heard outside the focus on this educational material encounter.

Critical making might not require the emergence of a material voice, but it certainly allows for it. That is why it serves as a useful educational reference to apply critical practice to vital media. For individual makers, a critical making practice can indeed lead to a recentering. To return to Perner-Wilson's work: she notes such a refocusing in her own practice. Reflecting on her own material practices, she highlights four key points: (1) she is not in the lead during the material practice but "following most of the time"; (2) the "materials contain the stories we use them to tell," and these stories are not "ours"; (3) she still struggles with the "transitioning between the abstract and the tangible"; and (4) her material-centered work "is not about engaging in an extreme material-lead adventure as much as it is departing from social frameworks of value" (Cole and Perner-Wilson 2019, 110). Her reflections mirror many of the points raised in this chapter. Recognizing the active role of materials shifts the concept of agency. The needs and activities of the material itself surface. For humans, these material encounters turn into "complicated improvisations" wherein activities are shared in a fluent collaboration that does not see a single dominant partner. Such an embrace of materials is a shift away from "social frameworks of value," which used to be the focus of interest.

In Perner-Wilson's case, this often collaborative work includes studying crafters (Buechley and Perner-Wilson 2012) as well as material experimentation (Mellis et al. 2013) and the development of new materials (Honnet et al. 2020) and tools (see the *Ohm-Hook*, figure 4.3). The work largely rejects the use of prefabricated kits and remains exploratory. These explorations remain material focused and support a kind of hybrid craft. Indeed, given the expertise Perner-Wilson has acquired over the years of this practice, she can be described as an expert crafter in this new domain. At the same time, her work has also been educational, as many of the projects are documented in detail, and she readily shares the practices and results of her explorations. She not only embeds digital technology into material practices but also shares the emerging practices online and through countless workshops. In that way, the explorative engagement reaches further and wider. When we add this focus on practice and on the emerging material objects to critical making, we arrive at an educational take of vital media. It can build on established pathways, such as Hertz's five steps, but it adds a focus on the object, its material agencies, and the human practices that speak to them. During the research, design, implementation, and deployment, we also need to ask: Did we follow the material's qualities? Is the object centered on a sustainable balance with its milieu? Did we abuse materials to make a social commentary? Or did we allow the limitations and abilities of the materials to inform the process? And to what new material variations did this lead?

Tracing the connections to objects, materials, and practices mirrors the efforts of others to help us look at the world through the eyes of nonhuman critters (Haraway 2016) or alien objects (Bogost 2012). But the purpose here is not to shift the focus to the Other, the "material," as the new center replacing the human one, either. Vital media are still mediation action networks, and they are still distinct from other actions because of the formation of human expression. This remains as integral to an educational approach as the inclusion of nonhuman factors. Vital media are not posthuman but they include the same constructivist material encounter that stimulates human learning processes in critical making. Vital media extend this to add more emphasis to the voice of the object and materials, which are elevated to that of an equal partner in the emerging practice. That means, that the results are not mere prototypes that amplify the creative mind of the human, but they are things in their own right.

Because vital media are inconclusive, critical approaches within them can be better centered on educational practices than on set models. Critical making as an educational project embraces materials and making and serves as a useful reference. Departing from the largely social agenda of critical makers, such as Ratto and Hertz, vital media depend on the voices of those materials and include them into the educational pattern. Instead of limiting critical making to an exploration, we are tasked with allowing the materials to come forward not as means to an educational end but as operators themselves. This can even include a step away from the social frameworks. Consequently, we have to give the emerging object results their own right to emerge. This making is still "critical" but includes nonhuman participants. The next section will discuss one example how the voices of the material can add to a critical process.

Broken Vessels

The task was to build a guitar synthesizer, a device that would combine synthesized sound generation with the kind of tangible controls that are typical for guitars. With a background in math, Amit Zoran worked on the algorithms to crack this challenge for a company in his home country, Israel. But he sensed a conflict with the premise of the project. "We come from nature and two guitars will have different DNA. Because the DNA, you cannot clone it, you cannot synthesize it" (pers. conv., November 25, 2019). Following this task, Zoran moved to Boston and worked on related projects, dealing with sound, guitars, digital fabrication, and hybrid instruments. The question of this hybrid nature followed him throughout this work, and the challenge of what can or cannot be synthesized would stay with him for years to come.

Zoran's family had reached Israel from Poland and Russia. Their story was part of the Zionist movement in an emerging state's history with the atrocities of World War II and their own identity as Ashkenazi Jews very much intertwined. The grandfather worked as a pianist in Poland; the father became an engineer in Israel; Zoran himself developed a background in engineering as well as industrial design. This included his turn to mathematics that prepared him for the aforementioned synthesizer guitar. The combination of disciplines and approaches prepared him for his next step. He moved to Boston and joined the MIT Media Lab. There, he continued his studies first toward a master's (which still dealt with guitars), then a PhD, and eventually a postdoctoral position. Along that trajectory, his work turned more and more toward craft and toward questions battling with the "DNA" of materials and objects. "I am a terrible musician and in being honest with myself, I understood that it's not about music for me it's about the craft. It's about the materiality and the tangible qualities and the difference between industrial design and traditional craft and what both of them have to say about engineering and then I started to investigate into hybrids" (Zoran, pers. conv., November 25, 2019). To follow that interest, he worked with such crafters as Marco Coppiardi, a violin maker trained in Cremona, the city of Stradivari and Guarneri. Coppiardi focused on traditional luthier techniques, following maker practices that date back to the seventeenth-century masters. This level of detail in traditional practice only deepened Zoran's interest in materials and the way we might engage with them.

When he left Tel Aviv to move to Boston, Zoran brought along a ceramic bowl made by his cousin. This cousin had been a budding ceramic artist but had tragically passed away the year before Zoran left for the United States. Part functional piece, part reminder of a lost and beloved family member, the bowl had traveled with Zoran and was now used in his Boston apartment. One day, a cleaner accidentally broke it. "This bowl had a unique texture and was not perfectly round; one could easily see it was a handmade object, a unique artifact that would not be confused with another. The bowl embodied my memories of him making it. It represented great emotional value, associated with deep family connections and important events. In 2010, this bowl was accidentally broken by a visitor to my house. The visitor suggested paying for the bowl. Of course I refused; there is no price that can restore a memory. The original meanings embodied by the bowl were irrevocably changed" (Zoran and Buechley 2010, 5).

The bowl did not "work" anymore, but the loss of functionality was as much part of this—now broken—piece as was its place within the family history. The story of the object, the collaboration with it, had continued ever since its physical making. Through the travel across continents and through the accident, the DNA of the object had developed another strand. The destruction was not the end of the object's story. Building on his work on craft and digital fabrication, Zoran decided to reengage, to continue to collaborate, with the broken bowl and explore the new condition. In that way, the destruction itself became a core element of a hybrid design approach that would combine craft, materials, and digital fabrication techniques.

An object's destruction is the ultimate risk but, by definition, always part of its nature from the crafter's construction process to our daily use of it to a visiting cleaner's encounter with it. Thus, the goal could not be a reversal of the destruction. Instead, Zoran had to integrate that destruction in the shared and evolving story of the bowl and help to realize it as a new part of the object's qualities. Zoran went ahead to reconstruct the broken bowl using 3D printed materials that would connect the remaining shards. The bowl remained as an object but changed its functionality. It could not hold liquids anymore, for example. However, it still holds the memories, the traces of a cousin forming it, as well as the signs of the destruction. The object continues to unfold, and it retains its altered voice.

This relationship of original and refabricated became the basis for a new piece of hybrid reassemblage. For a follow-up project, Zoran designed a vase in the 3D modeling program Rhino and used this model to fabricate the molds needed to cast three vases needed for the project. The three vases were cast, fired, and glazed. Of those vases two were deliberately smashed into pieces. The broken pieces were reassembled by Zoran in combination with 3D-printed parts that filled out the missing sections and refitted them into the remainders of the ceramic vases as seen in figure 4.9. "The resulting 'restored' vase functions as a memorial—a memorial that, for the maker,

Figure 4.9
Hybrid reassemblage: glazed ceramic, SLS nylon element, epoxy glue and black spray paint. Used by permission of Amit Zoran.

retains traces of the object's entire lifecycle: construction, destruction and restoration; a memorial of a beloved object and its breakage. The new one-of-a-kind piece acknowledges the original, the act that destroyed it and the process of restoring it" (Zoran and Buechley 2010, 5). In that way, function is extended, even emphasized through the addition of digital fabrication. A connection between an "extreme interpretation" of risk in material handling, destruction, and digital design and fabrication emerged.

The result can only be described as a *relational piece*, not between humans but across humans, practices, and materials. These hybrid vases are not means to an end, but they quite literally embody the tensions of an argument on destruction and reassemblage. "So when I look at it now, the damage is more important than the artifact. The damage is part of my full perspective on craft and on humanity and creativity. So you can call it a damage. You can also call it unpredictability or imperfection" (Zoran, pers. conv., November 25, 2019). Imperfection, signs of damage, difficulties, and even pain are spread across practices, objects, people, and every body's history.

The vases are embodiments of a damage that relates back to Zoran's initial struggles with synthesizing musical instruments. "We are synthesizing the perfect violin, the perfect guitar, but they're never perfect and they're getting damaged while you're using them. But what does it mean to get damaged? They are getting also value. They are changing and the mark of time and the mark of the atmosphere, the environment, and the way you use them is been downloaded into the artifact and become to be part of the artifact. So, by damaging the artifact you give it meaning and you make the bond between the user, or the musician . . . to the artifact itself" (Zoran, pers. conv., November 25, 2019).

Traditional craft has repeatedly been emphasized as a way to recall or reconnect. Adamson (2013, 184) talks of craft as "memory work," as a way to engage with a "trauma" caused by modernism. The "damaging" impact of the Industrial Revolution and how craft enthusiasm emerges as a response to it is part of craft scholarship and history. We see comparable societal impacts unfolding today, an era sometimes termed the "third Industrial Revolution," in the way that digital fabrication might affect and reshape practices. But Zoran's work is not conceptual art history. Unlike Adamson, who weighs Pye's concept of "risk" and Morris's praise for "imperfection" against each other, Zoran implements both rooted within the realms of personal fabrication and digital media. He follows not a theoretical but a deeply personal-material way. His is an encounter through practice and material engagement, where a look at "damages" and "imperfections" is not only about risk to the object or the person, "but it's about the investigation and the relationship with the impossible, with the unknown, with the difficult, with the missing element and how you define yourself along this negative space of creativity while you are evolving to be the person you are" (Zoran, pers. conv., November 25, 2019).

The hybrid reassemblages of the vases offer a story of imperfection in material, crafter, design, objects, tools, practices, us. Imperfection is the goal. Zoran went on to write a *Manifest for Digital Imperfection* (2016, 26) to call for an imperfection "that can engage the designer in an ongoing performative search for balance between the controllable and the uncontrollable in design and nature." The vases are one example of an imperfect and ongoing relationship and the emerging practices. As such, they are an example for vital media engagement that builds on the struggles involved in ongoing relating. Like Beuys's trees, the dance of *Subway* participants, or Tool's cups, Zoran's broken vessels do not and cannot conclude, but their broken bodies keep on emerging and the maker participates in that through practice.

Leaving Craft

A player of a video game will learn optimized play methods not only as tactical strategies but also as physical actions. For example, players might learn through hours of game play where to plant a trap in a 3D environment to fortify one's dungeon or which equipment to choose when embarking on a specific quest. But they may also learn tacit components from routinely pressed key combinations to input patterns that optimize the virtual avatar's performance or a virtual race car's maneuvering. As long as interaction with a computer requires directed muscular engagement, it is never disembodied. The challenge is what to make of this embodiment. This is the practice of collaborating with digital machines, but too often those tacit components lack specificity in their material realization. We might control a virtual car, avatar, or spaceship—all with the same game controller. This does not exclude digital media, such as video games, from being poetic or part of a constructive material practice. But it highlights their limitations in the current media conditions. It might also explain the increasingly detailed richness that game controllers have developed over time to capture human input, from touch pads, to analog buttons, to motion sensing, to voice and body tracking. The need is clear, but its realization in interaction and media design remains a challenge as long as these materials are only seen as tools without their own valuable voices.

Simply turning to material counterparts and maker practices does not resolve these limitations automatically either. A "maker culture" that focuses on tools or on kits that remain unclear in their functionality might mediate the operator away from the material at hand just as a standardized game system does. One might be able to download a 3D model from a distant server and feed it into a remote 3D printer to produce a desired shape. But this does not accrue to a holistic material encounter. It does not compare to Zoran's hybrid reassemblage process. Much like a packaged tourism tour through carefully selected showcase pieces, the exploration of the material universe is in danger of turning into prepacked approaches and heavily regulated means of operations. There is little room to improvise or to critically subvert. In their attempt to make every maker kit as effective as possible, designers are tempted to reduce the critical exploration to guarantee a successful functional outcome. But as the discussion above has shown, without shared critical exploration and collaboration, without failure and risk, the connection to the material remains fractured and underdeveloped. Craft does not make such an encounter easier, but it emphasizes the necessary challenge and provides examples of how to address it. Turning to craft does not provide a solution to the challenges of mediation and technological encounters. Craft itself is a challenge, a complication. But it offers clear reference points and practices for tackling this challenge. These help us in the ongoing balancing act of weighing human and nonhuman collaborations.

For starters, they include a focus on needs as driving forces for all partners involved—not only the human. The beginning of this chapter offered a range of definitions of craft as being both material construction and conceptual imagination. Such building differs from the abstractions of unspecific or disembodied encounters often provided by digital formats. The working definition of craft as cocreative material practices that are based on needs and further individuation has shown alternatives through a range of examples. Craft provides theories as well as practices to inform and shape open intra-active collaboration.

Collaboration, here, is a *poīēsis* of human as well as nonhuman contributors. This allows for a foregrounding of materials and objects as actively contributing things in and of themselves. They allow us to become together. The three phases of encounter, exploration, and collaboration helped to structure this interdependent becoming. However, they do not conclude it. There is no solution to the challenge of becoming together but only an emerging dialogue. This dialogue might reach from the hands of the potter to the lips of the customer as it deals with the trauma of war and destruction (as discussed in Ehren Tool's work); it might be a long-winding exploration into novel material combinations and practices (as discussed in Hannah Perner-Wilson work), a playful mixed media encounter (as discussed in Colin Stricklin's work), or a hybrid reassemblage (as discussed in Amit Zoran's work). All these projects show forms of vital media at work and materials working "hand-in-stuff" with humans. Neither of them is concluding this encounter. Tool's cups will operate as media telling about the horrors of war for generations to come. Perner-Wilson's instructional videos help to educate a wide range of budding enthusiasts in wearable technologies. The artifacts of Stricklin's game and Zoran's assemblies remain open, challenging, ready for more exploration, continuing their stories. This is why the argument turned away from mastery over the material to critical making as an educational approach with materials. Instead of skill as an achievement, the focus shifts to collaboration as an educational exercise. This is the picture of an ever-expanding exploration and of ongoing shared critical practices.

A principle that emerges at the heart of this recentering is *care*, a concept that has become increasingly important across disciplines. María Puig de la Bellacasa explores the notion of care in a more-than-human world. It does not come as a surprise that Bellacasa (2017) arrives at care not as a final state but a dynamic balancing exercise, much like the recentering of humans encountering nonhumans that shaped this chapter. Care is not a settling into a perfect state, but it can destabilize and disrupt the status quo. "The letting go of the controlling power of causal and binary explanation comes with an immersion in the messy world of concerns. Being in the things we plunge into unsettled gatherings; rather than observe them from a bridge, we inhabit the realm of more than human politics" (Bellacasa 2017, 33). Care creates trouble in a "messy world" and continues the shift toward a focus on relations "of more than human" conditions. Bellacasa makes it perfectly clear that the hierarchies and dynamics of these relations need constant correction. The corresponding term might be *reliance*, wherein care is not optional but a necessity that we have to practice as we recognize our shared coexistence. This is covered in the notion of shared needs, but Bellacasa's call for disruption through care and its openness invite another area of craft research.

What has been termed by Anne Wilson as "sloppy craft" stands as one example for how practitioners willingly leave processes unfinished and foreground these open, disruptive, and critical relations (Paterson and Surette 2015). "Sloppy" does not mean "unskilled," even though its aesthetics often toy with such an affiliation. It steps beyond the focus on optimized objects that present mastery of a craft through flawless technical production. Instead, pieces can emphasize a material rawness. That does not mean that the makers lack skills, but skill is applied to present a "sloppy" or "raw" piece that allows the agency of the material to find its own voice and open the piece to further exploration. These effects are visible in some of Josh Faught's fabric pieces. Threads dangle loosely, and fabrics hang unevenly as they are pulled down by unequal weight distributions. It was an encounter with Faught's work that inspired Wilson's original coining of the term "sloppy craft."

Faught is an expertly trained artist and maker tackling personal, as much as social-political, and material histories. His practice often starts from the materials themselves, their assembly becoming a narrative. They are woven, crocheted, and dyed by hand, and they often emphasize the differences of these processes and how they affect the numerous materials.

Faught's practices bring materials' voices to the forefront through distinctive material specificity. Materials are often combined from multiple sources, including wood, paper, fabric, and various paints and dyes in *Triage* (see figure 4.10). Their arrangement and careful construction bring forth both, the voice of the artist and that of the materials. Risk, here, includes the actions of materials as they operate through their qualities upon each other, not smoothed out by human mastery to combine them but retaining their different agencies. The specific way one fabric might be heavier than its neighboring ones and how this affects the overall structure of a piece "speaks up" instead of being corrected to some level of skillful perfection.

Figure 4.10
Josh Faught, *Triage* (2009). Hemp, nail polish, spray paint, indigo, logwood, toilet paper, pins, books, plaster, yarn, handmade wooden sign, denim, and gloves 203.2 × 304.8 cm. Used by permission of Josh Faught.

We find the same approach in the hybrid works of Perner-Wilson's exploration of materials that, through her skills, can offer new functionalities and in Amit Zoran's digital and hybrid reassemblages. Critical making with materials might replace a focus on mastery with an ongoing educational project, but the notion of sloppy craft shows that at the high end of the spectrum, even the skilled crafter finds ways to help and articulate the voice of the material. This includes a pathway toward a more-than-human form of care into possible media design. It asks the designer in turn to provide a carefully crafted network of actions that do not invite a participant to care "for" the environment but "with" it, to engage in an intra-active collaboration that has to realize the actions and emergent voices of all partners. Care, here, is shared growth, and craft provides possible pathways to support it. Through this support, craft took us to the material heart of vital media making.

In between performance and craft, the alternating and joining voices of cognitive participants and nonsentient materials have been discussed. So far, these have only covered an individual's encounter with material collaborators. Encounter, exploration, and collaboration were only traced in relation to one active human involved. The next step looks beyond the individual and searches for forms in which such a material dialogue can unfold on a larger social basis.

5 Digital Folk

The closing chapter turns from the individual encounter to vital media in a wider social context. The story of Cat Mazza's *Nike Blanket Petition* offers one glimpse into material social criticism. It introduces the main theme of reclaiming fields through vital media practices. In her example, this includes reclaiming industrialized icons through shared maker practices.

Following this lead into communal media activities, we turn to folk studies, a field that connects many of the key elements outlined so far. Along that path, Dorson and Burrison help to outline main threads and key qualities of folklife. These ensure that the turn to folk remains a critical stance that provides informative approaches to communal material culture and not an alternative or counternarrative to the digital era.

Folk pottery traditions in North Georgia serve as historic examples to deepen these critical and historical categories. These craft practices trace back to the environmental conditions as much as the historic-social changes over time. Even though the conditions change, the continuities among these folk approaches remain clear.

These qualities are condensed into three points: a focus on lived material culture, variation within tradition, and community-based practices. Each of them relates back to the traditional folklife as well as serving as a critical means in the digital age. In that way, folk provides the means for a communal rebalance of media and material engagement.

Craftivism provides a path to follow this rebalancing act, and it helps to identify criteria and mechanisms pertaining to how digital media and traditional handcraft combine toward a larger reclaim of material culture. Inspired by Betsy Greer's work, we will explore these relations as they emphasize the combination of crafted material practice with digital media forms. The effect empowers the maker and calls for a turn away from the computer as media and back to the computer as a tool.

The concluding declaration of dependence summarizes the key concepts toward a vital media ecology beyond humancentric terms. It reconnects back to Beuys's idea of the social sculpture to call for a focus on the many interdependencies encountered in the discussion throughout and their value for media design.

Nike Blanket Petition

In 1971, Carolyn Davidson was studying art at Portland State University and needed money for art supplies. By chance, a teacher overheard her complaints in the hallway and mentioned that he might need some design work for a start-up he was working on. Davidson came up with a dynamic curving logo for Blue Ribbon Sports and charged thirty-five dollars for the design. The company changed its name to Nike, the logo earned itself a name, "swoosh," and went on to become one of the most iconic brand logos in existence. Or so the story goes.

The narratives around the Nike logo are manifold and complicated. It is one story to dissect the logo's visual appeal and impact, how it distanced itself from predecessors and competitors, such as the three stripes of Adidas, and informed later designs, such as the vector logo by Reebok (Goldman and Papson 1998). It is another story to realize what the logo came to stand for. As Davidson later noted in an interview: "When I see my design in everyday life today, it's a little surreal and strange. While I'm proud of what I did, in some way I see it as just another design. It was Phil [Knight, the founder of Nike] and the employees at Nike that turned the business into what it was. If they didn't have the savvy, it would have been just another drawing" (Davidson and Rovell 2016). As a "savvy" company, Nike became enormously successful. The logo became highly visible. And the two became ever more intertwined. So close, that in some of their marketing strategies, the logo is presented without the name of the company. It was that logo that adorned Andre Agassi's hat when he won Wimbledon. It was in black and red on Michael Jordan's sneakers that were initially banned by the NBA as breaking uniform code. That logo was pushed via massive advertisement campaigns onto every platform available to form an increasingly valuable brand identity. It was part of campaigns that were revolutionary in their concept, design, and scope.

But the logo also became an icon for the production conditions of the very objects it was meant to sell. Nike attracted criticism as news surfaced about exploitative practices in sweatshops that manufactured their apparel. Critical reporting published condemning details about workers' conditions (Greenhouse 1997), which were met by more public statements from Nike. The narrative turned into legal challenges in 1997, when the consumer activist Marc Krasky sued Nike for false advertising. Nike countered this by arguing that advertising was protected as free speech. The dispute led to an argument about the role of free speech at large, which went through different courts and almost made it to the Supreme Court in 2003. Shortly after, it ended in a settlement in which Nike paid $1.5 million to the Fair Labor Association (FLA). Very much part of this public debate, the logo had become a sign of success as well as of exploitation.

Taking up the call for better working conditions for sweatshop workers, microRevolt, a collective founded by Cat Mazza hosted the *Nike Blanket Petition*. The project was active from 2003 to 2008, but thanks to its tangible media design, it continues to spread its message to this day. To criticize the exploitative practices of Nike, the project turned to the swoosh logo. The *Nike Blanket Petition* consists of multiple components: first, an online petition demanding humane working conditions for workers that manufacture goods:

> **NIKE BLANKET PETITION**
>
> I, the undersigned citizen [full name, city/state, country] declare that clothes for sale in my community, produced in the U.S. or abroad, should be manufactured in accordance with established International Labor Organization conventions and monitored by independent auditors that are neither funded nor founded by the corporations accused of labor exploitation. Corporations using subcontracted labor have a responsibility to ensure that their products must be produced in a safe and healthy workplace, without child or forced labor. Corporations using subcontracted labor also have a responsibility to ensure that workers are paid a living wage, as opposed to the legal minimum wage, and workers have the freedom to associate and form independent unions. (microRevolt, n.a.)

Instead of merely signing the petition as a digital file, microRevolt recreated the central swoosh logo as a crafted piece of knitwork assembled from many smaller patches. A photograph of the growing piece of knitwork was put online as a visual reference. As the petition was launched online and the signatures started to come in, each underwriter of the petition signed it in relation to this physical swoosh logo and its online image. The underwriters' names were associated with individual spots on the crafted logo and could be parsed in an online visualization of the blanket. This web-based visualization showed the growing blanket and allowed visitors to the site to roll their cursor over the blanket image and discover who had signed the petition already. Each coordinate of the online image triggered a different pop-up of a signature. In that way, the crafted logo blanket served as the anchor for the digital canvas that embedded signatures in an online visualization targeting the logo just as the growing list of underwriters for the petition targeted the company and its policies. Together, the project emerges as its own kind of hybrid assemblage. "There seemed to be this potential using the web for organizing, but also this other added element that I saw, which was like the pleasure of making. So this idea of doing something that we enjoy—crafting—as a way of taking down something that I saw it as important to resist against" (Mazza, pers. conv., November 20, 2020).

The concept of a shared blanket harkens back to the role of "need" in craft. The fundamental value of a blanket is immediately clear to anybody who encounters it. "Covering" is a foundational quality mentioned by Risatti in his discussion of craft. Creating an actual blanket builds on basic human needs, much like the fair working conditions. Mazza recognized a loss and disconnect caused by a culture of consumption and wanted to ask "how we make things that we need to cover ourselves" (Mazza, pers. conv., November 20, 2020). In her design of the activities surrounding the project, commercial consumerist logic faces off with a basic need-based craft approach. Another form of critiquing Nike's exploitative work practices was that supporters were asked to crochet a four-by-four-inch square and mail it to microRevolt. Making one's own piece for such a cover invited contributors to take a stand through their own craft practices. Contributors provided for the blanket's extension as a covering familiar item, while also voicing their opposition to sweatshops and exploitation. Additional squares were produced in local workshops and at craft and activist events (see figure 5.1). The individual

Figure 5.1
Collaborative work on the Nike blanket (left); final assembly of the full Nike blanket (right). Courtesy of microRevolt; used by permission of microRevolt.

squares were assembled into a tapestry-like frame surrounding the central Nike logo blanket that microRevolt had constructed. The combined outcome was a blanket, fifteen by five and a half feet, that was ultimately assembled in the winter of 2007–2008 and that still travels to exhibitions today to tell its story (see figure 5.2). Each square represents a participant's support for the petition. It also forms a material construct that reframes the Nike swoosh logo as it becomes the center of the critique regarding exploitative working practices. But the blanket was more than a visual reference or metaphorical banner. For Mazza, it embodied the very nature of the semiotic regime that it engaged with:

> If you look at Fair Isle knitting from the Northern coast of Scotland, or if you look at the Chullo hats from Peru: the patterns in there are embedded imagery from their region. So they have like native birds or views from the Andes in their knit patterns. . . . I'm so fascinated by that. It's like compressed designs from the landscape. And so like that was the other comment with the sewing was like: No, our landscape is consumption. It's this, like, this is the semiotic regime, is like this brand. And that's why it wasn't Nike swoosh with an X through it, because that would be something else. (Mazza, pers. conv., November 20, 2020)

Much like the traditional knit patterns of historical folk traditions, the Nike blanket inscribed elements of daily life into its physicality, into its fabric. Only this time, those were not the patterns of regional birds or landscapes. These were the patterns of global marketing, production, and consumption. Making it part of the activist project, the *Nike Blanket Petition*

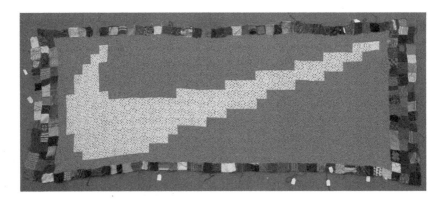

Figure 5.2
The *Nike Blanket Petition*; crochet squares assembled to a representation of the swoosh logo (2003–2008). Courtesy of microRevolt; used by permission of microRevolt.

reclaimed this mechanism for those who participated in the blanket construction. "If you look at the symbolic power of a brand in any way as infiltrating our landscape, or as like this oppressive, having some sort of oppressive, place in our global economy, in our climate in our workplace, etc. It was a way to take it back. Like we own that. We made a piece of it" (Mazza, pers. conv., November 20, 2020).

The crafted blanket became a statement in itself. It continued to tell its story as it traveled to different exhibitions once the process of its production had concluded. I encountered the physical Nike blanket in the "Making Change: The Art and Craft of Activism" show at the Museum of Design Atlanta (MODA) a decade later, and it still served as an evocative example of activism through craft practices. It still "worked."

Digital practices were important in more than one way for the project. Before the *Nike Blanket Petition*, Mazza and microRevolt had released a free open-sourced digital tool, *knitPro*. *KnitPro* still works today and allows crafters to create customized knit patterns for their own use with the help of an app. Crafters can submit images, which will be rasterized and turned into knit patterns by the application. Visitors to microRevolt's site would encounter both projects next to each other, and those who downloaded *knitPro* would also see the link to the Nike online petition. The tools for one's own making practices and the call for a critical application of those practices stood side by side, and the community of crafters interested in digital patterns met the activism against exploitation of any such craftwork. The central swoosh logo part of the Nike blanket itself followed a rough digital pattern that could have been generated by *knitPro*.

Over time, the digital parts of the project faced their own technological challenges. *KnitPro* remains usable, but following a piece Mazza did for the Hillary Clinton campaign, her servers were hacked, and the blog that housed her various projects crashed. Eventually, the Flash plug-in for the website, needed to visualize the mappings of signatures to the digital logo blanket, became defunct. Browsers moved away from that technology, and the visualization has never been updated. Today, the full functionality of the digital components can only be seen in online videos and still images. In many ways, the physical components of the project outlived the digital ones. Yet the project emerged from a hybrid practice that relied on traditional crafting and constructing of a physical piece as much as digital communication and dissemination via blogs, emails campaigns, and other outreach. The resulting petition is a multilayered message that consists of hybrid artifacts based on varying techniques from the digital Flash plug-in to the crafted yarn. They were initiated by a single person, Cat Mazza, but realized by a much larger group of people in the spirit of tactical media. The personal physical and the digital histories counter the industrial abstracted narrative of the Nike logo through handiwork and individual petitioning. As such, they present an example of reclaiming a media artifact for communal interests. Participants create their own version of the very logo they criticize, inscribing their own names onto that handcrafted cultural landscape and reclaiming it through a craft as much as a digital intervention. It is not anymore left to the savvy of the Nike executives to infuse Davidson's original design with new meaning. This new version, the Nike blanket as an embodied petition, tops the commercial construct through distributed social activism, and it does so via a combination of craft and digital technologies.

Entering Folk

So far, the key criteria of vital media have been applied only to the individual encounter and to how one person and the surrounding materials realize their interdependencies. Performance as a practice of mediation was described in the moment of individual production that hinges on the moment we create personal expression in collaboration with the material. It did not depend on the success of that creation with others. It was not yet important for our working definition whether personal performances were understood, appreciated, or even noticed. Instead, the focus was on the action to produce expression with an intent. Likewise, crafting presented a personal encounter with the material. It looked at the act of manipulation and the intimate encounter with the stuff at hand, not its community-building history, commercial viability, or emerging communities of practice. Both positions depend on their embeddedness in the world, and the practices were *based on the needs* of all partners. Through encountering, exploring, and collaborating, these practices further the individuation of all their participants.

Both key fields, performance and craft, make it clear that vital media are production driven. They are a shared becoming between cognitive and noncognizant partners. Vital media, those peculiar networks of activity that allow for the production of expression and material distinction in a balanced way, make it clear that we only emerge as unique beings through intra-active encounters with nonhuman materials.

Such an approach centers on individual encounters and helps to clarify the necessary moment of production. But it inherently comes with a range of limitations. Individuation and personal expression, distinct from anybody else's work, are trademarks of Western individualism. We might have balanced the individual with their surroundings through vital media, but the focus on one person cannot be the end of this argument. The encounters are always also part of social construction and communication. Markets, legal regulations, communities of practice, publics, and histories are all part of the discourse that vital media stand in. The concept of intra-action expands our involvement outward into the world of matter and material phenomena, but it also includes intra-human action. Will vital media hold up in societies and in communities of shared work? Or is it limited to the individualistic actions of a person trying to differentiate themselves from their group as they turn to the materials they work with?

This section draws from folklore studies and anthropology to expand the initial idea of vital media to wider social networks. It asks where and how this approach operates in society and the environment as it turns the focus on groups of people and their changing dynamics over time.

Turning to folk poses its own dangers. The term has been abused and instrumentalized before. Folk-based approaches have been cited to support evolutionism and racist thinking. They were tools feeding into the worst offspring of such ideologies. *Volk* was central in Nazi propaganda and served as a cultural underpinning for the horrendous terrors of the Third Reich (Dorson 1972). Local and folk traditions were used to boost nationalism, to justify political means, to draw opportunistic cultural divisions, or to justify plain exploitation or racism. When we turn to the study of folk culture, we need to remain aware that it is in constant need of critical revision itself. There is not a single folk culture that could be used as a "solution" to the challenges at hand.

Like craft and performance before, folk does not provide such a solution but a critical perspective. It is not *a* folk community that might resolve the balancing tasks a hand—it is the application of folk *as such* that informs the necessary work. The study of folklife offers a focus on qualities that speak to vital media as a constructive balancing act, but this does not resolve social inequalities or exploitative practices. Just as we reject the myth of a return to the "good life" through a craft revivalist stance, this argument cannot simply turn to "better communities" via folk. The goal will be to apply folklife perspectives to explore larger social groups and their emergence in relation to the material world they live in. Often, these communities are rural. But instead of centering on the regional differences, the more important condition is their relationship to the surrounding material partners and how these relationships distinguish them. "What sets them apart, perhaps, is that they are less concerned with the larger society's values and structures than with working within their own understanding of the world. This approach sometimes reflects a cultural way of moving through the world that may seem odd or foreign to outsiders" (Congdon and Hallmark 2012, 7).

Dorson (1972) divides the study of folklife into four main threads, each of which can contain different types of media: *oral literature*, such as folktales and songs; *material culture*, which responds to "techniques, skills, recipes, and formulas transmitted across the generations" (2); *social folk custom*, with an "emphasis on group interaction rather than on individual skills and performances" (3), which includes such customs as marriage, rites of passage, festivals, folk medicine, and religious practices; and *performing folk arts*, such as dance, music, and drama. Based on those four core threads, he lays out a wide range of methodological and theoretical approaches to studying folklife. Notably, these threads of folklife contain the two main domains of performance and craft. It sees no gap between the material culture of local crafters and regional performing arts. Including the two core fields of performance and craft as separate material practices within the concept of folk provides an inclusive überdiscipline, a key for the attempted conclusion of a call to vital media practices into social practices.

If Dorson lays out a range of active formats in which folklife can materialize, then Burrison suggests qualities of folklife within them. He distinguishes between half a dozen criteria that outline folklife as

- *learned traditionally* (i.e., through informal personal teachings rather than through written or otherwise documented instruction)
- *community-shared* (i.e., working with "community resources, supporting group values while serving individual members' needs" (Burrison 2000, 29) which indicates a belonging to small groups with shared resources)
- *bringing the past into the present* (i.e., even modern folk practices and pieces build on a chain of past references and historical precedents)

- *flexible* (i.e., even though they connect to the past, folk traditions live through adaptation and alteration)
- *producing objects that are both useful and beautiful*
- *handmade* (i.e., in opposition to machine produced). (Burrison 2000, 29–36)

Burrison's qualities can apply to every thread outlined by Dorson. For example, folk dances are shared within communities and kept in the style of past traditions; so are religious customs, crafting techniques, and recipes. These qualities differentiate locally shared knowledge from DIY approaches and distinguish folk traditions from other emergent practices, such as self-taught art. They offer a first glimpse into what could be specifics of a digital folk idea. They also connect social to material conditions in a most fundamental way where community resources are material as well as social. Traditional folklife practices, such as hunting, fishing, or farming, are examples of such a shared reliance. This fabric provides an entry into a culture that incorporates a "living off the land" and practices that remain closely tied to regional material conditions. Local customs might be "flexible" but remain dependent on regional conditions in terms of both social and material fabric. A local dish not only depends on the shared knowledge within the community and how its recipe is carried on through the generations. It also depends on the availability of the ingredients in the region.

At the same time, innovation and tradition support each other. They remain intrinsically connected by "artists who both adhere to a particular cultural context and are individually innovative about the way they make their art" (Congdon and Hallmark 2012, 7). Changes still happen through personal expression by individuals, but they might depend on local conditions more than on conceptual stylistic forays. They remain "individually innovative" but also tied to the materials, histories, and shared resources at hand. Each of them requires its own time and space, following seasonal weather patterns and growth cycles as much as shared customs and traditions. Craft asks us to realize the different time frames of the materials and tools surrounding us. Folk asks us to recognize the socially shared temporalities of traditions and historical practices that are specific to a place.

> What makes all of this folk behavior, as folklorist define it, is learning process. The way that traditions are passed on within a group, from mouth to ear, from body to eye over the generations across time and space retaining the core element of the tradition, but evolving at the same time, changing at the same time, and developing variation. So, you know, one of the ways folklorist study a folk tale

for example, is to collect all of the variants that had been published or that are in archival collections in manuscript, and then attempt to use those, to reconstruct the history, the life history of that folk tale, to try to figure out where, and when it may have originated and in what form and how it has evolved since then, as it's moved into different countries and over different time periods. We do the same thing with material culture, material folklore. And so the basic concept is the same. The same principle for determining whether a pot or a building is a folk artifact applies to a song or a story in terms of understanding where the ideas are coming from. In the case of material culture, we're talking about the design of a building or the shape of a pot. We're talking about the materials and techniques of construction. So that's different from telling a traditional story or singing a traditional song, but even those have their parallels to material culture. (Burrison, pers. conv., September 14, 2018)

If we project these dependencies forward, then a version of digital folk has to connect to shared materials and practices as well as shared social resources and customs. Both are interdependent, and it is this interdependence that shapes distinctive features of different folk art and life. The social conditions are forever changing. This applies to rural communities as much as global citizens. A return to rural lifestyle in the age of overpopulation and urbanization is unfeasible and the turn to folk does not ask for one. Instead, it emphasizes that these interdependencies are not limited to rural communities. Mazza's *Nike Blanket Petition* involved an international community of crafters that came together over a shared concern as well as over a shared material artifact. Connecting local practice to material conditions is possible in urban as much as rural communities. Digital media can support qualities to hybrid materials and practices just as any other technology can. They can be learned traditionally through personal instruction and informal teaching, community-shared and working within the resources and value systems of a group, bring the past into the present and referencing traditional sources, they are highly flexible, can be both useful and beautiful, and even handmade.

To repeat: folk, like performance and craft before it, does not provide a conclusion. It has produced countless social and cultural difficulties and inequalities. The turn to principles of folklife is not a return to a better past condition. It cannot be used to hide social injustice, past or present. What it does offer is a lens to envision a possible relationship between larger social bodies and material conditions through shared practices and particular qualities, such as Burrison's and Dobson's. And, as we are living in the digital age, it allows us to ask what it may mean to "live off the land" when the land includes digital media. To tackle this question, we first turn to an example of local folk art that has managed to survive up to today yet has seen changes throughout its development, which might help to fill in the qualities and threads established in this first turn to folk.

Turning and Burning in North Georgia

Today, the outskirts of the Blue Ridge Mountains in the southern part of the Appalachia region stretch across the Carolinas and into the state of Georgia. Before European settlers arrived, the area was part of the "Enchanted Land" of the Cherokee Nation as it extended farther south to the Atlantic coastline. Signs of settlements have been documented through finds of early pottery across the coastal region as well as up the rivers. They date back to 2600 BC and are among the earliest documented pottery artifacts on the North American continent (Sassaman 1993). Tracing such artifacts is important because pottery marks one of the technological pivot points in humankind's development. The ability to store and prepare food opened up developmental opportunities for hunter-gatherer tribes and supported evolving farming communities. Emerging cultures built on technologies like these to change sociotechnological conditions. This emergence of ceramics as well as that of different social structures unfolded gradually over the Late Archaic Period (3000–1000 BC) on the North American continent. During this process, the potters of this region used particular materials and practices, such as pit firing and coil building. Later ceramics of the First Nations show the use of patterns and an emerging aesthetic as many of these potters used stamps and wooden paddles to make their wares more ornate. The distribution of the resulting First Nation pottery across the region varied. Its production depended on, among other things, the availability of materials as it spread along the Savannah River and the coastal regions (Sassaman 1993) where the necessary resources were available. But not unlike other pivotal technological developments, the history of pottery also unfolds in irregular, disjointed, and painful steps that center on sociohistorical development as much as on environmental factors.

European colonization, and with it the onslaught on the First Nations who lived in the region, followed its own pattern. At the time of large-scale European settlement in the first half of the 1800s, the region was home to the Cherokees bordering on the Creeks. At first, pottery traditions evolved during the initial period of contact between the tribes living in the region and the European settlers pushing into it. But in 1838 the Cherokee were forced out of the region in what ultimately led to the Trail of Tears. Their original pottery techniques almost died out before they were revived in the early 1900s (Duncan et al. 2007). Traditionally, pottery was a craft performed by Cherokee women and historical techniques were carried forward by them into today's practices (Duncan et al. 2007). Famous pottery families included the Bigmeat family, whose matriarch Charlotte Bigmeat is credited with continuing Cherokee pottery practices in the Qualla Boundary of the Eastern Band of the Cherokee—now in North Carolina. It is through continuing practices like this one that their lineage of ceramic art and production remains alive.

European settlers moved into the region largely from the north, bringing with them different pottery techniques and practices developed overseas. These included the use of the wheel to center and manipulate clay, as well as their own glazes, recipes, and different firing techniques. Abner Landrum adapted Chinese stoneware pottery techniques in the early 1800s, probably from his own library and through experimentation. The resulting stoneware techniques replaced the dominant lead-infused glazes with alkaline ones with notable health benefits. Landrum was a physician who must have seen the effects of lead poisoning in his patients. Thus, ceramic practices in the 1800s turned from earthenware to stoneware, a tougher and more durable material that needed higher firing temperatures.

The new techniques did not detach the settlers from the conditions of the land, though. Burrison traces the development of the folk pottery tradition not only in the movements of settlers from the northeast into the region but also along the Fall Line of the Piedmont Plateau that marks the North Georgia geological landscape. Along this Fall Line, increased deposits of stoneware were located and supported the settlement of potters. This "clay country" (Burrison 2010) was also fertile farmland, which in turn necessitated pottery to provide wares for the storage of produce. Pottery remained a need and clay a local resource that filled this demand. As Lanier Meaders, one of the celebrated folk potters of the region, noted about the role of the potter and ceramics: "Just about everything else at one time in this part of the country depended upon it. . . . You know, necessity rules everything" (Burrison 2008, 15). To provide for these needs, potters mined and prepared their own clay locally, mixed their own glazes using the materials available locally and those that came in via trade from the coast, and fired their ware in wood-fired kilns. Even today, some folk potters still gather clay in the region, and the red Georgia clay has its own qualities and aesthetics. Likewise, the recipes of glazes can be handed down across generations within families to this day. As a result, the development of European-style folk pottery in the region has been recognized as its own cultural heritage, distinct from other ceramic traditions, such as those in the US Southwest:

> So we've got the use of indigenous materials. Yes, we've also got ideas probably borrowed from as far away as China, via printed sources. And we've got the shapes of the pot by potters with training going back to England and Scotland, like the Landrum family in the Edgefield district. So, there are all these things coming together, it's not just the use of local materials. That's part of the story. But the settlement history of a given region is as important for understanding how, not just material folk culture, but all folklore arises in a given area. Who were the people who came into the area, introducing traditions from other parts of the country and from other countries? (Burrison, pers. conv., September 14, 2018)

What emerged was a distinct folk pottery culture that depended on environmental conditions, social changes, realized through individual practices. In South Carolina, Abner Landrum worked with a talented slave for one of his businesses. He would eventually become known as Dave the Potter, a unique crafter, who not only excelled in making large vessels but who also signed and marked his own pots with poetry in a time when literacy among slaves was illegal. Material folk culture cannot hide such social injustices, but it constructs and reflects the culture it is part of, socially as well as environmentally. It is in the individual practice, the construction of the pots in the hands of Dave the Potter, that the dialogue between material and personal realization comes to life. The resulting pot would be sold and used locally to fill a given need. But the same pot shows that marking an object can be a political act, not in the programmatic sense but in the more fundamental matter of establishing one's presence in a society within this circle of need-based production-expression.

For the arriving settlers, the longer growing seasons of the South supported a largely agricultural economy and the landscape led to scattered family farm businesses that often had to operate as independent units. Living here meant a continued "frontier experience" within a "region timelag" (Burrison [1983] 2008). But it also included making something you personally would find beautiful and expressive based on local education and local resources and knowledge. "Folk art, mean something different to everybody, I'm sure. But to me it is without formal training. Just what you're naturally inclined and able to do just out here in the middle of the woods and middle of nowhere and you're making something beautiful out of nothing" (Mike Craven, current folk potter, pers. conv., June 28, 2019). Informal training was provided by family members or via apprenticeship. "There is a heritage going on in the practices. The past is being used and recycled if you will, as a resource by the next generation to carry on a family tradition, a community-based tradition. You know, in the old world normally to create something tangible you didn't just invent the design and the way to make it in a vacuum. You learned those things from others" (Burrison, pers. conv., September 14, 2018). Training, like recipes, was handed down within the community and often within the family—leading to the famed "clay clans," defined not only by social bonds but also by material practice.

"Although there is some leeway for individual variation, it is replication, or conservative fidelity to the inherited tradition, that governs the folk potter rather than innovation or the desire to create a unique product. The pottery designs, then, were slowly refined as they were transmitted through the generations, becoming the shared property of families, communities, even regions of potters. It is as the embodiment of a collective tradition that we can best understand a piece of folk pottery" (Burrison [1983] 2008, 57). To this day, local use of specific clay mixtures, glazes, and forms remains distinct. Typical glazes of the Southeast, such as the alkaline "tobacco-spit" glaze, still differentiate local ceramics from other productions. Another typical element is the lack of ornamentation and how this somewhat changed with the shifting market conditions. Originally, the settlers' pottery produced in northern Georgia rarely showed any decorations but focused on functionality. Among the rare decorative features might be markers that indicated the capacity of a jug to hold a certain number of gallons. Over time the original need for the wares diminished as industrially produced containers replaced the necessity for ceramic storage. Glass bottles replaced the original whiskey jugs, for example, which had been a staple for local potters. Once, perishable wares were stored in large jugs that were left standing in cool streams. Now, refrigerators took over the task, and along the way, one can sense the aforementioned turn to Borgmann's commodities. Cheaper and faster industrial production of dishes changed the commercial validity of these original sectors of handmade folk pottery. One response of the local crafters was a shift to more ornamental pieces, such as regional face jugs, that were often produced not with a view toward function but aesthetics.

Today, conditions have changed even more as most regions are more accessible, and new communication channels allow folk potters to sell their work at local fairs as well as via online portals. To this day we can find multiple folk potteries operating in the region that are either part of or closely associated with the clay clans Burrison initially identified. Pottery as a distinct folk practice remains active and concentrated in that region, where it has been studied (Eaton [1937] 1973) and documented in the histories of leading folk pottery families, such as the Meaders (Rinzler and Sayers 1980) and the Hewells (Glassie 1999). The most comprehensive study of this continuing folk pottery tradition was conducted by Burrison (2000, [1983] 2008, 2010), who also assisted in curating the exhibition at the Folk Pottery Museum of Northern Georgia, which opened in 2006. The museum continues the history of ceramics in the region as an exhibition space as well as an educational space for the preservation of practices. It provides both a showcase exhibit and a center for workshops and outreach programs in the region. In these instances, folk is active in both production as well as in reflection on the practice.

To survive, folk traditions change and adjust constantly. Today, potters rarely dig for their own clay in Georgia even though some of it is still locally sourced. Traditional wood-firing is still practiced but has become the exception due to the higher cost of wood compared to gas or electric kilns. Many folk potters still have traditional kick wheels but use electric wheels in their daily work. These changes were only logical as Steve Turpin, one of the practicing potters of the region puts them into context with changing limitations and necessary adjustments: "They [past folk potters] work with what they had and I work with what I have" (Turpin, pers. conv., June 28, 2019). Within this mindset, moving from a kick wheel to an electric wheel is not breaking a tradition but a form of continuity. It is not a disregard of past traditions but a continuation of the ethos of a folk potter.

These practices of contemporary folk culture show two qualities: an ongoing *lived material culture* and a *variation within tradition* that carries this culture into the future. Although some traditional techniques and materials have been superseded, the notion of folk pottery remains and grows. Mike Craven, another practicing folk potter, supports this further: "But we all have to change, and we love the knowledge and technology. And you're crazy if you don't use it. You have to use it to your advantage. And try not to forget the roots of where we came from" (Craven, pers. conv., June 28, 2019). Maybe that is why both contemporary folk potters cited here have a deep interest in keeping this practice alive through teaching and as part of the community. "I hope people remember me for my pottery. But what I want to be remembered for is: if I knew something and I could help you—I help. That's where the community comes from" (Turpin, pers. conv., June 28, 2019). Material production and creative realization become the center of social construction even in modern times. They manifest through teaching and community building. These *community-based practices* are a third criterion to capture modern folk practice.

Dorson supplied the initial breakdown into four fields of folk study along oral literature, material culture, social folk custom, and performing folk art. In addition, Burrison provided specific qualities across these threads: learned traditionally, community-shared, past into present, flexible, useful and beautiful objects, and handmade. These categories distinguish how a grown community of folk potters differs from hobby and studio potters. This kind of differentiation might highlight some opportunities for vital media production. Folk potters might have changed traditions to adjust to modern needs and to continue their way of life. But studio and hobby potters never had those traditions to start with as they emerged from an entirely different community and tradition that often centers on techniques, technologies, or educational practices. The specifics of the region in the Deep South make those components concrete for the communities and the pottery practice at hand. They include historical and environmental context, defined as part of a larger history, utilitarian in design and appearance, limited in themes, reliant on specific regional materials in local clays and glazes, continued practices taught within communities, and need-based production embedded in community and family traditions. They situate and specify the principal categories in the region and community. The (necessarily fragmented) discussion of how these folk traditions evolved over time to still operate today were traced to the criteria of ongoing lived material culture, variation within tradition, and community-based practices. They build on the much wider frameworks of Dorson and Burrison to turn the focus on more current phenomena in folk.

Some of these recurring components discussed here have been central for design and specifically interaction design (such as context and materiality); others are more challenging for design particularly on the industrial level (such as regional specificity, small-scale practices, traditional learning processes). How do we apply these folk principles to media design? Lived material culture, variation within tradition, and community-based practices help to distinguish folk practices and offer criteria in the turn to digital folk. The following outlines these criteria further and connects these traditional folk practices to the initial picture of hybrid and craft-activist approaches to suggest ways in which we can design for a digital folk practice.

Digital Folk Practices

Folklife's variation within tradition and its embeddedness in material conditions and the "naturally inclined" practices that emerge from material culture as a living community-based practice provide a triangle of references. These three cornerstones can be summarized as follows.

Lived material culture combines a dependency on shared needs as well as a close connection to materials sourced in the region. In the case of the regional Georgia folk pottery, this is visible even today in terms of practices (e.g., the continued use of traditional glazes), resources (e.g., the mining of local clay), and aesthetics (e.g., the persistence of farm-based themes and motives). Materials and regional conditions support particular practices that grow and change over time. While a range of media and interaction design projects aim to bridge the digital and the physical through new interfaces, folklife never produced the abstracted divide between local tradition and material engagement in the first place. This supports a shared material culture that remains aware of material dependencies and includes them in its practices. Pottery evolved where clay or other material necessities, such as water or wood, were available, and the resulting wares were meant to support regional life and needs.

Variation within tradition supports a social evolution of such practices and material developments. Family, local, and regional traditions provide continuity and they slow down development as they fit any change into an existing tradition. They might change slowly and maintain individual expression within a longer, regional aesthetic, but change itself remains an integral part. Folk practices have to adjust to new needs, technologies, and social upheavals. They do so not by design but through gradual emerging shifts of the shared milieu. They remain, to use Burrison's criteria, flexible as they bring the past into the present. In contrast, the fields of media and interaction design have traditionally emphasized prototyping and fast creative variation over the grinding slowness of such traditional practices. Variations in new technologies are often praised as global standardized solutions, and new technologies are distributed with worldwide impact, little local footing, and compressed development and release cycles that can be dictated by commercial factors beyond seasonal or regional changes. Folk traditions follow a slower pathway where iteration cycles still vary but are bound to local materials, community practices, and established aesthetics and themes. They call for a slowing down within the locale. This provides the necessary time for reflection in relation to the over-time changes of materials and conditions, whether this refers to the speed of drying clay or repetitive knitting patterns. Variation still happens in the individual encounter with the material, but that variation is not bound to a sped-up construction process. Changes in these practices are not aimed at industrial standards. Instead, they are geared toward a continuation of traditional approaches. Henri Lefebvre evokes the notion of rhythm as a deeply relational principle. It offers change through repetitive cycles, such as seasons, as well as linear gradual progression, which "generally originates from human and social activities, and particularly from the movements [*gestes*] of work" (Lefebvre 2004, 90). Change and tradition are interdependent and continuous, but most importantly, they are evolutionary, not revolutionary.

Community-based practices ground practice in locally shared knowledge. Knowledge and identity formation combine, for example, in self-organized pottery fairs or the sharing of family recipes for glaze mixes. These communities of potters "stay family" through such mechanisms that generate shared resources, interdependent practices, and communal needs. The goal is not only to maintain the practice but also to continue a social tradition and a communal coherence. This contrasts with some principles of distributed knowledge and educational media design that often emphasize online communication open to everyone yet with limited dependencies between the watching audiences. A folk-inspired approach might not be as open or accessible, but it might provide cohesion through its community focus. Such community-based practices do not favor the exceptional individual but they value sharing of expertise and social interdependencies. The differences become especially clear in different educational approaches. Folk craft explicitly rejects self-taught approaches and depends on local—often family- or tribe-based—education that is delivered from person to person. Typical teaching methods of maker culture or media education differ from such approaches. Their goals of lowering entry thresholds for engagement, emphasis on distributed knowledge and mediated communication, and simplifying complicated construction techniques through personal production technologies can be highly efficient in introducing newcomers to their techniques. But they do not follow a digital folk logic when they detach from community-based approaches. This does not devalue novel fabrication approaches. The fact that integrated approaches are possible has been shown in many local maker spaces as well as in the use of new technologies in established communities. Yet it emphasizes that the functional social fabric of those spaces has to remain in the foreground. Folk demands a rethinking of any claim of universality in maker culture and replaces it with the unique conditions on-site and within the community. In folk-inspired cases, making techniques focus on the support of a distinct shared identity as much as on the construction itself. As Steve Turpin noted: "I hope people remember me for my pottery. But what I want to be remembered for is: if I knew something and I could help you—I help. That's where the community comes from" (pers. conv., June 28, 2019). The story of the practice is the story of the person, often the story of their families and neighbors, too. It cannot be the story of everyone everywhere. Such local and personal specificity might restrict the sharing of knowledge and reject assistive interfaces aimed at lowering access thresholds if those interfaces interfere with the social needs. But they emphasize education and practice as forms of a locally specific social dialogue. The social activity of teaching and being taught are as foundational as the practice of making and the resulting objects. It follows that a fabrication lab cannot be defined by the types of fabrication tools that have been installed. It has to define itself from the community of participants and their material partners.

These three key elements emerged from a look at local folk traditions in the southeastern United States, but they are not limited to a particular region. They are connected to the foundational role of materials in any discussion of media activities. Within design thinking and research, they connect to theories of autonomous design that emphasize autonomy and community. Basing his work on Latin American designers and activists, Arturo Escobar (2018, 16) proposes a radical rethinking of design toward an ontological design approach: "Together with the recrafting of communal forms of knowing, being, and doing, these notions—*autonomía* and *comunalidad*—and their associated practices may be seen as laying the ground for a new design thought with and within communities." His more politically based design mirrors many of the points that are key to the folk-based argument above. It argues for more sustainable practices through bioregionalism and builds on Maturana and Varela's (1992) work on autopoietic organization. Maturana and Varela originally developed this concept from a discussion of reproductive powers and detail it through the example of cell development, among others. For Escobar, key features of this organization apply to the Indigenous communities of Latin America. He emphasizes the concept of territory as shorthand for necessary relation systems that support communities, self-sustaining communal units, and their close connection to the material conditions of the land they live on. *Autonomía* and *comunalidad* remain the central themes throughout, and for Escobar (2018, 175; emphasis in the original), "*autonomía is a theory and practice of interexistence and interbeing, a design for the pluriverse.*" It should apply to situations "in which communities relate to each other and others (say, the State) through structural coupling while preserving the community's autopoiesis" (Escobar 2018, 173). His overall argument is far more driven by a social-political agenda than by a material-based one. It emphasizes community-based practices and proposes a turn away from oppressive state and global systems. Adjustments are still tied to communal tradition. This argument calls for its own version of variation within tradition as one of the descriptions of autopoiesis is "changing traditions traditionally" (Escobar 2018, 173). In Escobar's case, the role of materials and things is discussed in relation to emergent social alternatives, but it is clear that these communities only come into being through close relations with their environments. The directionality of lived material culture might differ, but it remains an important factor.

There is reason to hope that already existing communities as well as emerging ones might offer the necessary conditions for vital media to thrive, that we can learn from these communities and expand from them outward. However, this is not the argument of the book at hand. First, it would be an anthropological project and different from this book's focus on media design. Second, this logic might suggest that particular folk or regional communities might provide an available solution to the challenge of balance in vital media. This would be self-contradictory. The argument for vital media does not demand that we copy from any single community or art or craft but to keep on learning from all of them. This is an argument to support ongoing mechanisms, not for final solutions.

For the time being, it is necessary to discuss how the three key elements of lived material culture, variation within tradition, and community-based practices can be realized in the digital media age and look for opportunities for how they might support and foster vital media practices. How then can such folk-based criteria apply to the digital media domain? After all, Mazza's *Nike Blanket Petition* project might have been inspired by the logic of local practices and tactical media, but it was initiated by a single person. It did not emerge from a local tradition but collected parts from all over the world into one critical piece. It references not a gradually grown functional form but is centered on a logo designed by a single designer: a logo that was chosen because it conflicts with sustainable local resources or practices, one that emphasizes global commerce. Yet these are precisely the conditions that shape the digital media landscape today. What the *Nike Blanket Petition* delivers is a critique of practices that are perceived as gone awry from social and environmental sustainability. It does not resolve them, but it is a corrective that reorients media design and individual as well as social engagement. This reorientation is a vital media effort that answers to the elements of digital folk. This reorientation includes collaborative workshops (to create parts of the blanket or assemble them), sharing techniques and technologies (like the *knitPro* software), and a reclaiming of the landscape that global branding has left us with (through the blanket as well as the online petition). Instead of negating the dominance of global media practices that have shifted the balance to the companies, or the economic factors which control production, marketing, design, and commerce, the *Nike Blanket Petition* meets them head-on. The rebalancing effort builds on existing conditions and critiques them instead of avoiding them through a turn to a past ideal. The project does not reject the logo but adopts it and reframes it. It literally puts it into a frame made by many protesters in the form of the physical blanket and transforms the logo into a digital signature page in the online visualization.

The call for digital folk is not a Luddite return to predigital handiwork but a revision of the conditions of collaborations between human and human as well as human and nonhuman in the world we share right now. That is why it does not demand fewer digital components. It does, however, require an inclusive design to balance cognizant and noncognizant cocreation in a landscape that inherently operates via established material as well as digital qualities. In Mazza's case, it is a corrective of practices with the help of handiwork. This corrective itself depends on digital media but turns to folk qualities to come to life. How can these correctives be summarized and supported? Are there models for such a turn to alternative practices that go beyond single projects and that could apply to a wider range of vital media through performance- and craft-related means?

Criteria for Digital Folk

Calling for a focus on lived material culture, variation within tradition, and community-based practices does not reject any form of technology per se. It does not open up an alternative ideal to escape to. Instead, turning to digital folk is largely a process of correction. One story for a related rearrangement is that of Betsy Greer. Encouraged by the Riot Grrrl movement, Greer turned to crafting and DIY. Eventually, she wanted to volunteer for work with senior citizens and was looking for contacts. This led her to try a knitting circle:

> I found somebody at work that knew someone that had a knitting circle, and then all of a sudden it was a space where you walk in and everyone's super nice. Everyone's teaching you different things. And someone sat me down and my friend down and said, "This is what you do." And then they left, and then someone else came up and said, "Oh no, no, you're doing it wrong. Do it this way." They laughed. Someone else came up and said, "Oh, you met . . . you messed up. This is how you fix it." . . . There are all these different interactions with people based on their own personal histories with craft because they all learned from their grandmothers or from their mothers. (Greer, pers. conv., June 2, 2018)

Greer realized her fondness for knitting as part of her social volunteering. Somewhat surprised by this realization, she asked how such an old-fashioned, slow practice could become so important for her in an age of instant gratification. The social encounter with others was an important component, but so was the effect on herself. "There was something about the way they both allowed me to bring myself into the present, to just hang out and get comfortable (instead of trying to relive the past or jump ahead to the future) that was absolutely brilliant" (Greer 2008, 1–2). She experienced the craftwork as calming, connecting, as well as inspiring for herself. Instead of providing her services to these communities, it became clear that these groups of traditional makers were helping her. Finding and supporting a community through shared practices points back to Escobar's call for *comunalidad*.

Inspired by the discoveries of these traditional but lively practices, Greer went on to popularize the term *craftivism*. Craftivism combines craft and activism into material practices that might support resistance, discovery, or empowerment. As a term, *craftivism* took off quickly, but it is only one thread in a wider movement within craft, which combines social means with material practice. This includes notions of a "handmade nation" (Levine and Heimerl 2008) or Gabriel Craig's "church of craft" (Bryan-Wilson 2012). They share a subversion of a hypercapitalistic system, its media ecology, and ultimately the social narratives that it forces upon its users. These subversions manifest through a focus on material practices:

> Craftifesto
> The Power is in your hands!
> Craft is powerful
> We want to show the depth & breadth of the crafting world. Anything you want you can probably get from a person in your own community.
> Craft is personal
> To know that something is made by hand, by someone who cares that you like it, makes that object much more enjoyable.
> Craft is political
> We're trying to change the world. We want everyone to rethink corporate culture & consumerism.
> Craft is possible
> Everybody can create something!! (Carlton and Cooper in Levine and Heimerl 2008, xx)

Crafting and its objects, their performative making, and their distribution are used as political media activities. Parts of this movement grew from a slowed-down counterculture favoring handicraft over mass production and self-made products over commercialized goods. That does not mean that it rejects the digital. Craftivism itself was publicized by Betsy Greer on her blog, via newsletters and podcasts, and through her textile work, writings, and presentations. Craftivism, much like Mazza's *Nike Blanket Petition*, uses digital media as integral parts next to physical craft approaches. The correction is the reintegration of material crafting practices, not the abdication from digital ones. One example for this is the *You Are So Very Beautiful* project.

The *You Are So Very Beautiful* project was initiated by Greer in 2015 and combined online media with traditional cross-stitching. Its Twitter hashtag #yasvb (for *You Are So Very Beautiful*) goes hand in hand with traditional handcraft practices, such as sewing, cross-stitching, or quilting. Both are means of the project's main theme, which focuses on affirmation and support. The project's website asked participants to create affirming messages in the form of small cross-stitched signs, place them in public spaces, take a picture of these handmade signs on location, and post them online with the hashtag #yasvb (see figure 5.3 for one example piece). Signs are built individually or in workshops, and drops of handmade signs can be unique individual events as well as coordinated efforts at craft fairs or even internationally organized occasions. It is a critique of the "world where the media constantly tries to tell us what is in, out, cool, passé. Every day we have to fight to remember that we are enough just as we are, we are beautiful just as we are" (Greer 2017).

The local, almost hidden nature of the handmade sign merges with online social media mechanisms, and the project works through the differences and combination of the two. The affirmation is personal, local, handmade, and hidden. At the same time, it translates through social media into a globally shared, easily identified digital version that reaches far wider audiences. It utilizes digital mechanisms as it can be searched and followed via online services. Digital media are not rejected; they remain integral to the project, but the handmade portions offer a critique of the overwhelming speed and communication demands of the dominant mediascape. It uses the same digital media that distribute images of perfection to remind us that "we are enough just as we are." The project reframes this media use and reshapes the stories told. Through a correction of the activities that make up this craftivist media effort, a participant reintegrates handiwork in the media mix to support personal and social affirmation.

"I could write a tweet, but that's a lot different than making a stitch—holding a stitch piece in my hand that I made myself. It's that tangible

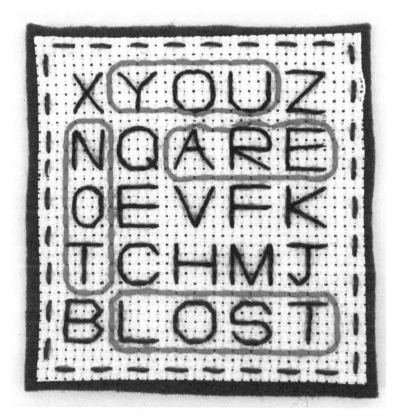

Figure 5.3
Example cross-stitch for #yasvb by Olisa Corcoran (2016). Photo by Olisa Corcoran. Used by permission of Olisa Corcoran.

object of something that I made from scratch. [It] is essentially what I find really poignant and important" (Greer, pers. conv., June 2, 2018). *You Are So Very Beautiful* builds on a variation within tradition, as it turns to the long-established maker technique of cross-stitching and crafting. It realizes lived material culture by supporting personal and community affirmation through the material as well as digital practice. Each piece is hybrid as it consists of a crafted making and digital sharing effort. It builds on community-based practices not only through workshops and public events but also in its inherent goal to critique and correct the society we live in. It supports communal self-affirmation through a piece that "I made myself" and that encourages and supports those who encounter it. It is a hyperlocal as well as distributed craft-activist performance based on shared personal needs, which uses local encounters through the nature of the drop of a handcrafted piece in a particular neighborhood in combination with the digital social media integration.

Finding one's voice through craft, encouraging others to find it, and making it heard are central concerns for Greer. She adds another view of craft to the ones that opened this chapter. Greer recognizes the functionality and need-based nature of craft but further emphasizes the empowerment that comes from the making itself. This making is for Greer the construction of one's voice, one that does not require perfection. "I think that everybody has the power to come up with their own projects and their own things. And I don't think everyone believes in themselves enough to think that they can do it" (pers. conv., June 2, 2018). To that end, the physical construction process helps to produce the necessary self-confidence and self-reliance. It provides a channel to develop an opinion, to take a stand, and to voice it outward through material means. This does not require a participant to be a master of a trade, nor does it depend on a thing well made, but it emphasizes the power of individual searching for a personal form of material expression. This personal expression is much more important than aesthetics or levels of flawless craftsmanship in the piece.

Speaking about the Tool cup we met at the beginning of the craft section, Greer clearly sees the agency within the cup, its function, the choices embedded in it. She notices the traces of the practice, and through them, she senses Tool's decisions. "You start thinking about how there were choices in that cup" (Greer, pers. conv., June 2, 2018). From this, she reads the cup as a statement at work and continuously collaborating with its owner. But it is also an encouragement that invites us to embrace its imperfections and unique features as personal choices. "That cup gives you then permission to start making your own things or at least think about the possibility of making something. Because I think a big part of being makers in general is giving permission to other people to do stuff and to fail and to make things" (pers. conv., June 2, 2018). The handcrafted item provides a supporting call to action to whomever it may encounter. It is this voice of the material that is also part of a project such as #yasvb as it combines material crafting and digital media to do, fail, grow, and gradually develop one's voice. This works only in a combination of craft practices with digital ones, such as blogs, online forums, Twitter campaigns, or data visualizations. As was the case with the Georgia folk potters, sharing these practices becomes ever more important and opens up the social construction of the communities that drew Greer into craftivism in the first place.

The second effect of this combined approach concerns the idea of the computer as an experiential tool and object itself. The tactile immediacy of the handmade object is often juxtaposed to the abstracted digital component, but the two do not have to be counterpoints. As individuals learn the tools and the practices to find their voice in such craft-based techniques, and as materials contribute to the emergent expression as active partners, we can assign the same role to computers. The *Making Core Memory* project (Rosner 2018; Rosner et al. 2018) builds on the handcrafted computer core memories of early Apollo space missions. These included interwoven wires that were carefully arranged by craftswomen to generate the necessary logic structures—literally weaving the code structure into the hardware. *Making Core Memory* built on this work, recognized the women who performed this crafting as code making and carries their stories and practices forward into workshop settings for today. In structured workshop events, today's participants are invited to create their own memory core components and assemble them in relation to the historical ones (see figure 5.4). The project

Figure 5.4
Making Core Memory quilt showing historical core elements in relation to workshop-created quilt parts. Used by permission of Daniela Rosner, University of Washington.

rightfully shines a light on unfair gender biases and working conditions, it foregrounds craft-based construction processes and embodied making, and it emphasizes the entangled nature of this computing form between abstract and manual operations. It is also a piece that works with the materiality of the computer itself. The computing logic is not detached from the crafted one as external force. Instead, the crafted object embodies the computational logic through its design and functionality.

This shifts the role of the computer away from that of a multimedia machine toward that of an experiential device. In *Making Core Memory*, computing is part of material culture in a way that embraces the intermateriality of hands and wires, humans and machines, physical and abstracted computational logic. The computer is no longer a detached mediating machine or a threshold object into an "other" form of representation, but it returns to its much earlier state of an actual tool and emerging object itself. It is through direct hands-on engagement that we can build new meanings, take part in object differentiation, and enact individualization.

For Rosner (2018, 11), the project works as a form of "critical speculations" in design that encourages us to think about design differently, in a way that is "both activist and investigative, personal and culturally situated, responsive and responsible." Rosner adapts this approach from Saidiya Hartman, who initially saw critical fabulations as a way to contest existing points of view and the authority of a single perspective, as well as to offer a space for the voices of the unheard. In Hartman's case, these grew from the unheard voices of slaves that historically remained unrecorded and would remain impersonal and excluded from the archives. She argues that the goal is not to give a slave woman a voice but to construct a piece that draws the core challenge of these missing voices into the present. "The outcome of this method is a 'recombinant narrative,' which 'loops the strands' of incommensurate accounts and which weaves present, past, and future in retelling the girl's story and in narrating the time of slavery as our present" (Hartman 2008, 12). At the end of this kind of critical fabulation, there is no new conclusive historical account, but the work on such a fabulation opens up critical reflection on the absence of one. It invites speculation on alternatives and requires continuous engagement moving forward.

Making Core Memory emphasizes issues of labor, inequality, and bias as the designers apply feminist approaches to technology and practices of making. These are necessary corrections that often already include a turn to material concerns, such as questions of environmental justice or sustainability. For better or worse, vital media differ in their point of departure. They do not center on social matters but intradependent material ones. Where the two align in their conclusions is that neither critical fabulations nor the turn to digital folk claim to deliver solutions. Both are correctives and means for course alignments. Both actively bring these corrections into the present with a clear connection to their sociomaterial context. They differ in their argumentation and in their weighing of operational components, but both reject a solutionism that would promise a new, somehow stable state from where we can proceed. Both are human-scaled approaches that build around changes in our consciousness, and one can sense the underlying educational project that they require. This can clash with more technologically driven proposals for futures.

Digital folk is not the call for lifelike robotics and general AI systems. It is not a principle for the internet of things, nor does it inherently require machine learning. Instead, it is a call to foreground localized and community-based as well as material-cultural conditions into the way we construct our expression through digital and nondigital media. If we apply vital media to the internet-of-things paradigm, then it is much more about the "things" and how they coshape who we (humans and nonhumans) are than about the "internet" (the commercial and social phenomenon) as driving factor. This is not a rejection of digital formats as it readily integrates them into its hybrid network of activities. It is a readjustment to consider them part of our local traditions, our personal histories, our lived and enacted realities, our sociomaterial cultures. As *Making Core Memories* illustrates, we might confront the computing logic itself anew as a material encounter. But no adjustment will ever arrive at a stable end point. Making vital media is a constant act of correction that does not comply to any single blueprint. Vital media are the suggested means for us to build the necessary ways of meaning making, storytelling, and fabulations to keep on correcting.

Building Forward

The search for the vital media ideal has wound itself through performance and craft into digital folk. Along this journey were encounters with many projects and approaches that realized the action to produce expression with an intent as much as the cocreative material practices that are based on needs and further individuation. These projects were not taken from any single media technology or period. They reached from ancient pottery traditions to performance art of the 1970s and 1980s to augmented reality applications and modern hybrid maker practices. Each of these examples offered its own take of the central question of balance and intra-active becoming. They all included cognitive contributions as much as noncognizant ones and served as references to develop the ideal of vital media as networks of activities that realize material agency, incorporate human cognitive abilities, concretize the object, and support the individuation of the human. The scale differed wildly, ranging from citywide social sculptures to individual crafting experiences. This range serves a purpose. Vital media as an ideal is a correction that can and should be applied to many media technologies and on every scale. That does not mean that it applies equally to all. It certainly conflicts with some approaches more than others and will operate easier in some technologies than others. For example, any format, such as most virtual reality approaches, that aims to separate the body and the role of objects and materials from the cognitive experience will struggle with such an ideal. It looks critically at many dominant digital media formats, from social media to video games to streaming platforms.

At the same time, this struggle is at the heart of the necessary balancing. Vital media are an ideal and rely on instable systems. Differences are feeding this system and instability is not the problem. The problem is that media designers have to try harder to balance the mediation of materials and humans. The examples showed some of this reach in Mazza's *Nike Blanket Petition* as well as in Tool's cups and Quitmeyer and Ansari's AR dance project *Subway*. The shorter examples often offered a reinterpretation of a piece within the vital media call, as was the case with *Tweenbot* or *Blendie*.

Vital media remain an offspring of the human condition, but they are not humancentric. They support human participants in their individuation as much as objects in their concretization and the shared ever-changing milieu they bring forward together. Instead of revolutionary digital innovation or social upheaval, these vital media are based on continuous evolutionary forces. They move in time spans that are bound to the physical collaborators as well as the social fabric that they are part of. They are process based, critical, inconclusive, and ever emergent. They reject dreams of humancentric dominance and emphasize intra-active dependencies. They do not look back for solutions but for relations and aim to bring forth such relations into future critical responses. This is not a position that is popular in today's mediascapes. The argument of this book often stands in contrast with the logics used to drive future media designs that emphasize speed, comfort, human experiences, commercial viability, or technological innovation. At the same time, vital media do not reject digital technologies. They suggest an inclusive corrective. Digital folk provides the framing for the proposed dependencies. Digital folk does not paint an alternative of escapism but instead draws in necessary correctives to materialize and enter the conversation.

It is in the nature of an evolution-based model that there is no conclusive end. But the continuous act of balancing between human and nonhuman contributions offers stabilizing forces in itself. Leaving the center empty allows us to realize the interdependencies as steadying forces. It also helps us to recognize moments where these interdependencies are neglected and our media are out of whack. The mechanisms we use to position ourselves and our activities are too often misaligned, and vital media suggest corrections that will reach from environmental to social adjustments depending on the project and the format. Those conditions change all the time and vital media are continuously balancing, open-ended, and unfinished. They are not prototypes for a defined larger project. They reject the idea of finalization and thus cannot be prototypes for any solution at the end of any process. Each object—each crafted patch, hybrid assemblage, mediated dance—is a valid contribution along the path, but vital media do not tell us where that path should lead. This distinguishes vital media from other design approaches that aim at tactical impact or direct activism. The vital media concept developed in this book is not "good," but it is necessary.

The first chapter laid out parts of the problem. It seems frighteningly clear that a human-centered world view is heading toward catastrophic results. Yet the media we use to address this challenge are still following a humancentric logic. We use the tools that led to the slide into our demise as our only means to address it. This mismatch was a key concern driving this book. Media design needs to change. We need to revise how we think and engage with media because we have to revise how we think and engage with the world around us. We need to look for different media assemblies that are sustainable in themselves. This calls for a continuous balancing act.

The second chapter established the idea of vital media as a means for this balancing act. Realizing the agency of materials and the specific cognitive contributions of humans alike, it sketched out the contours of vital media based on evolutionary principles. These focus on the individual yet interdependent development of all participants as they depend on shared growth. Vital media are expressive and productive at the same time because they realize human cognition and material agency as interdependent. With such a skeletal map at hand, the argument turned to two central practices to develop the concept of media for humans as well as materials: performance and craft.

The third chapter picked up the idea of performative activity as constitutive, that is, as reality constructing. Performance was defined as the production of expression with intent, not a representation of some earlier produced text. It directly responds to the need of cognitive beings to express and construct themselves forward. The examples and discussion in this chapter dealt with that construction as it happens in collaboration with the surrounding partners and targeted digital media within these networks. It rooted them in a material foundation with a look at Beuys's *7000 Oaks* project as one example of material agency realized and reaching beyond human communication, activity, and time frame. The same interdependencies were at work in the excursion into puppetry as a material-centered performance practice. We found them further in helpless robots or digital dances powered by cell phone applications. Without denying humans their role in the production process, this section allowed us to turn away from a human-centered design principle and toward a material interdependency while still holding on to a particular human contribution.

The fourth chapter established the productive role of media in synchronous material production. Humans and nonhumans, cognitive and non-cognizant beings, construct each other through intra-action, through a shared calling into correspondence that provides a "response-ability" (Haraway 2016). This is constructive both in a performative as well as a material sense and led to a turn to craft. Stepping through encounter, exploration, and collaboration, this construction turned social as well as material. Vital media combine needs and opportunities of all partners involved, and craft practices provide clear examples for such a balancing. Tool's cups are vessels for liquids to drink as much as storytelling objects that depend on the hands of the potter as much as on the clay they are formed from. If the performance chapter corrected the dominance of the human actor, then the craft section emphasized the growth of the material's contribution. Here, in a procedural learning through critical improvisation in close proximity to the materials, vital media practices emerge as forming all partners involved. This can take the form of a hybrid reassemblage—like in the broken vessels of Zoran—as well as a material-based video game or hybrid craft practice. Focusing on the emergence of these relations led to a look at care as a more-than-human principle that calls us to assist the rise of material voices.

Performance and craft are not presented as solutions. Neither solves the challenge at hand, yet both help in describing the necessary corrections as we turn away from humancentric media without attempting a posthuman scenario. The argument itself remains trapped in an ongoing balancing act that includes humans as one of the active ingredients. This last chapter expanded from the individual encounter to a wider social one. Continuing this back-and-forth, digital folk does not claim to provide an alternative conclusion either. Instead, it picked up the critical threads and continued them toward corrective action. Starting with Mazza's *Nike Blanket Petition* and expanding it into the mechanisms of craftivism and critical fabulations, it suggests a social-material reclaiming of media activities. Media, those networks of activities that allow both human expression and material realization, cannot return to a past folklife. But folk helps us to realign the mechanisms of how we encounter, express ourselves, listen, and support the voice of material agencies within larger social structures. Lived material culture, variation within tradition, and community-based practices offered criteria for a social-material turn to digital folk, which culminated in ongoing corrections exemplified by craftivism.

This results in a declaration of dependence: a dependence on material, biological, and cognitive needs. This dependency is environmental as much as social because it sees both inherently tied to each other in the evolutionary project they are part of. Because of the ongoing changes that define evolution, the underlying needs are ever evolving, and they cannot be resolved. That is why vital media do not provide a unified theory to solve the challenge that material media throw at us.

In Zoran's broken vessels, we sense the story of the clay as much as the personal tribulations of the maker. In Tool's cups, we find the concerns of the maker embedded in the markings. The need to drink and Tool's urgency to share his experiences both materialize in the nature of the cup, which continues its object story. The object is given space, range, and occasion to "speak up." Greer's #yasvb project lives on Instagram as much as local discoveries that surprise and support us through possible physical encounters. The nature of a drop of handcrafted affirmations in a neighborhood includes the promise of these objects to be ready for discovery, to surprise the finder as much as the crafter who planted it. This promise is digital by nature but a central component of the project. As we make, encounter, and share the objects over time, their voices and contributions assemble. These encounters can grow into ongoing collaborations. They might stretch over a couple of minutes, as in the drying Hydrostone of Stricklin's *Primal Clay*, or months, as in a collection of the *Nike Blanket Petition* pieces, or many years, as in the still growing trees of the *7000 Oaks* project; or they can be condensed to single frames in an animation, as in the *Subway* project. The range of examples used throughout this book aimed to emphasize that vital media do not refer to a single technology but an ideal to applied to all media technologies. Each one will ultimately have to respond in their own way. Still, the turns from human-centered expression to inter-dependent material conversations, to educational and critical encounters, to reclaiming shared needs and mediascapes outline mechanisms to work with. They just do not assemble into a single framework or even a manifesto to conclude. Instead, vital media call for an embrace of the balancing practice itself. This sets out the tasks for the creators of vital media. The role of the media designer has to be less that of the expert providing solutions and more that of one who expertly hands over practices to materials and humans alike, so they may thrive together. Thankfully, there are already examples that embody this task well executed—and some of them were discussed in this book as they propose their own ideas of how to balance agencies and needs. The most difficult area to address is to foster the concept of digital folk.

Expanding the idea of the maker to a multitude is one way to embrace digital folk. Instead of countless individuals creating separate content components, digital folk builds on communal and material trajectories. Communal practices, workshops, teaching, and learning practices within a social setting are central to projects that seem to carry the idea of the social sculpture into the digital age. There is unbound potential in this. But this is not an argument for a particular community of practice nor one about particular ecological necessities that might lead to "better" futures. Vital media do not bind themselves to a single topic but instead set out to reframe media into networks of activities that *inherently* realize such challenges through a balance of material agencies and objects next to those of humans and communities. Vital media have a hard time to forcing their logic into a commercially viable concept as they are not market driven. The concept might indeed be principally nonviable in a purely capitalistic marketplace. But it invites us to rebalance the way we build stories and explore our own futures and thoughts through the necessary inclusion of the materials we are embedded in.

What vital media require is human change as a necessary correction and response to material change that has already happened. There is no alternative to behavioral change, and that is what the call for the vital media ideal ultimately adds up to. We depend on it.

References

Aarseth, Espen J. 1997. *Cybertext: Perspectives on Ergodic Literature.* Baltimore: Johns Hopkins University Press.

Abrahamson, Dor, and Robb Lindgren. 2014. "Embodiment and Embodied Design." In *The Cambridge Handbook of the Learning Sciences,* edited by R. K. Sawyer: 358–376. Cambridge: Cambridge University Press.

Adamson, Glenn. 2007. *Thinking through Craft.* New York: Berg.

Adamson, Glenn. 2013. *The Invention of Craft.* London: Bloomsbury.

Alfoldy, Sandra, ed. 2010. *NeoCraft: Modernity and the Crafts.* Halifax: Press of the Nova Scotia College of Art and Design.

Anderson, Chris. 2012. *Makers: The New Industrial Revolution.* New York: Random House.

Ansari, Ava, Media Farzin, and Molly Kleiman. 2012. "The Back Room in Conversation with Media Farzin." Six Degrees (blog), New Museum, April 9, 2012. https://www.newmuseum.org/blog/view/the-back-room-in-conversation-with-media-farzin.

Auslander, Phil. 2008. *Liveness: Performance in a Mediatized Culture.* 2nd ed. London: Routledge.

Austin, J. L. 1962. *How to Do Things with Words.* Oxford: Clarendon.

Barad, Karen. 1999. "Agential Realism: Feminist Interventions in Understanding Scientific Practices." In *The Science Studies Reader,* edited by Mario Biagioli, 1–11. New York: Routledge.

Barad, Karen. 2003. "Posthumanist Performativity: Toward an Understanding of How Matter Comes to Matter." *Signs* 28 (3): 801–831.

Barad, Karen. 2007. *Meeting the Universe Halfway. Quantum Physics and the Entanglement of Matter and Meaning.* Durham, NC: Duke University Press.

Barlow, John Perry. 1996. "A Declaration of the Independence of Cyberspace." Electronic Frontier Foundation. Accessed September 18, 2021. https://www.eff.org/cyberspace-independence.

Bell, John. 2014. "Playing with the Eternal Uncanny: The Persistent Life of Lifeless Objects." In *The Routledge Companion to Puppetry and Material Performance*, edited by Dassia N. Posner, Claudia Orenstein, and John Bell, 43–53. Florence: Routledge.

Benford, Steve, and Gabriella Giannachi. 2011. *Performing Mixed Reality*. Cambridge, MA: MIT Press.

Bennett, Jane. 2010. *Vibrant Matter. A Political Ecology of Things*. Durham, NC: Duke University Press.

Bergson, Henri. (1911) 1998. *Creative Evolution*. Translated by Arthur Mitchell. Mineola, NY: Dover.

Blumenthal, Eileen. 2005. *Puppetry: A World History*. New York: Harry N. Abrams.

Boal, Augusto. 2000. *Theater of the Oppressed: New Edition*. London: Pluto.

Boeing Company. 1967. *Lunar Orbiter 1: Photography*. Contractor report CR-847. Washington, DC: National Aeronautics and Space Administration. https://ntrs.nasa.gov/citations/19670023005.

Bogost, Ian. 2012. *Alien Phenomenology, or What It's Like to Be a Thing*. Minneapolis: University of Minnesota Press.

Bollmer, Grant. 2019. *Materialist Media Theory: An Introduction*. New York: Bloomsbury.

Bolter, Jay D., and Richard Grusin. 1999. *Remediation: Understanding New Media*. Cambridge, MA: MIT Press.

Bonanni, Leonardo, Amanda Parkes, and Hiroshi Ishii. 2008. "Future Craft: How Digital Media Is Transforming Product Design." In *Proceedings of CHI '08 Extended Abstracts on Human Factors in Computing Systems*, 2553–2564. New York: ACM.

Borgmann, Albert. 1984. *Technology and the Character of Contemporary Life*. Chicago: University of Chicago Press.

Borgmann, Albert. 1999. *Holding on to Reality: The Nature of Information at the Turn of the Millennium*. Chicago: University of Chicago Press.

Boud, David, Rosemary Keogh, and David Walker, eds. 1985. *Reflection: Turning Experience into Learning*. London: RoutledgeFalmer.

Bourriaud, Nicolas. 2002. *Relational Aesthetics*. Translated by Simon Pleasance and Fronza Woods. Dijon: Les Presse Du Reel.

Bryan-Wilson, Julia. 2012. "Body Craft: Preaching, Performance, and Process." In *40 under 40: Craft Futures*, edited by Nicholas R. Bell, 41–50. Washington, DC: Renwick Gallery of the Smithsonian American Art Museum.

Buechley, Leah, and Hannah Perner-Wilson. 2012. "Crafting Technology: Reimagining the Processes, Materials, and Cultures of Electronics." *ACM Transactions on Computer-Human Interaction* 19 (3): 1–21. http://doi.org/10.1145/2362364.2362369.

Bunnell, Katie. 2004. "Craft and Digital Technology." World Crafts Council, Metsovo, June 1, 2004.

Burrison, John. 2000. *Shaping Traditions: Folk Arts in a Changing South*. Athens: University of Georgia Press.

Burrison, John A. (1983) 2008. *Brothers in Clay. The Story of Georgia Folk Pottery*. Athens: University of Georgia Press.

Burrison, John A. 2010. *From Mud to Jug: The Folk Potters and Pottery of Northeast Georgia*. Athens: University of Georgia Press.

Butler, Judith. 1990. *Gender Trouble*. New York: Routledge.

Byers, Bruce K. 1977. *Destination Moon: A History of the Lunar Orbiter Program*. Washington, DC: NASA.

Carlton, Amy, and Cinnamon Cooper. 2008. "Craftifesto." In Faythe Levine and Cortney Heimerl, *Handmade Nation: The Rise of DIY, Art, Craft, and Design*, xx. New York: Princeton Architectural.

Caronia, Antonio, Janez Janša, and Domenico Quaranta, eds. 2014. *RE:akt! Reconstruction, Re-enactment, Re-reporting*. Brescia: LINK Editions.

Castells, Manuel. 1996. *The Information Age: Economy, Society, and Culture*. Vol. 1, *The Rise of the Network Society*. Oxford: Blackwell.

Causey, Matthew. 2006. *Theatre and Performance in Digital Culture: From Simulation to Embeddedness*. New York: Routledge.

Chapple, Freda, and Chiel Kattenbelt, eds. 2006. *Intermediality in Theatre and Performance*. Amsterdam: Rodopi.

Chatzichristodoulou, Maria, Janis Jefferies, and Rachel Zerihan, eds. 2009. *Interfaces of Performance*. Farnham: Ashgate.

Cole, David, and Hannah Perner-Wilson. 2019. "Getting Lost and Unlearning Certainty: Material Encounters in an Electronic Craft Practice." In *The Critical Makers Reader: (Un)Learning Technology*, edited by Loes Bogers and Letizia Chiappini, 107–126. Amsterdam: Institute of Network Cultures.

Congdon, Kristin G., and Kara Kelley Hallmark. 2012. *American Folk Art: A Regional Reference, Volume One*. Santa Barbara, CA: ABC-Clio.

Crocket, Dennis. 1999. *German Post-Expressionism: The Art of the Great Disorder 1918–1924*. University Park: Pennsylvania State University Press.

Csikszentmihalyi, Mihaly, and Eugene Rochberg-Halton. 1981. *The Meaning of Things: Domestic Symbols and the Self*. Cambridge: Cambridge University Press.

Cubitt, Sean. 2017. *Finite Media: Environmental Implications of Digital Technologies*. Durham, NC: Duke University Press.

Davidson, Carolyn, and Darren Rovell. 2016. "How a College Student Created one of Sport's Most Iconic Images." ESPN, June 17, 2016. https://www.espn.com/espn/story/_/id/16286876/how-college-student-carolyn-davidson-created-nike-swoosh.

Davis, Virginia. 1995. "William Morris and Indigo Discharge Printing." *Journal of William Morris Studies* 11 (3): 8–18.

Dewey, John. 1910. *How We Think*. Boston: D. C. Heath.

Dewey, John. (1938) 1997. *Experience and Education*. New York: Touchstone.

DiSalvo, Carl. 2012. *Adversarial Design*. Cambridge, MA: MIT Press.

Dixon, Steve. 2007. *Digital Performance: A History of New Media in Theater, Dance, Performance Art, and Installation*. Cambridge, MA: MIT Press.

Dobson, Kelly. 2004. "Blendie." In *Proceedings of the 5th Conference on Designing Interactive Systems: Processes, Practices, Methods, and Techniques*, 309. Cambridge: ACM.

Dobson, Kelly. 2008. "Kelly Dobson: Gel 2008 Talks." Gel 2008, New York, April 24–25, 2008.

Dormer, Peter. 1997. "Craft and the Turing Test of Practical Thinking." In *The Culture of Craft*, edited by Peter Dormer, 137–158. Manchester: Manchester University Press.

Dorson, Richard M., ed. 1972. *Folklore and Folklife: An Introduction*. Chicago: University of Chicago Press.

Dourish, Paul. 2017. *The Stuff of Bits: An Essay on the Materialities of Information*. Cambridge, MA: MIT Press.

Duncan, Barbara, Brett H. Riggs, Christopher B. Rodning, and Mickel Yantz. 2007. *People of One Fire*. Tahlequah, OK: Cherokee National Historical Society.

Dunne, Anthony, and Fiona Raby. 2001. *Design Noir: The Secret Life of Electronic Objects*. Basel: Birkhäuser.

Eaton, Allen H. (1937) 1973. *Handicrafts of the Southern Highlands*. New York: Dover.

Escobar, Arturo. 2018. *Designs for the Pluriverse: Radical Interdependence, Autonomy, and the Making of Worlds*. Durham, NC: Duke University Press.

Fischer-Lichte, Erika. 2008. *The Transformative Power of Performance: A New Aesthetics*. London: Routledge.

Foerster, Heinz von. 2002. *Understanding Understanding: Essays on Cybernetics and Cognition*. New York: Springer.

Fox, Paul. 2006. "Confronting Postwar Shame in Weimar Germany: Trauma, Heroism and the War Art of Otto Dix." *Oxford Art Journal* 29 (2): 247–267.

References

Fuller, Matthew. 2005. *Media Ecologies: Materialist Energies in Art and Technoculture.* Cambridge: MIT Press.

Galloway, Alexander. 2006. *Gaming: Essays on Algorithmic Culture.* Minneapolis: University of Minneapolis Press.

Galloway, Alexander, Eugene Thacker, and McKenzie Wark. 2014. *Excommunication: Three Inquiries in Media and Mediation.* Chicago: University of Chicago Press.

Gershenfeld, Neil. 2005. *Fab: The Coming Revolution on Your Desktop—from Personal Computers to Personal Fabrication.* New York: Basic Books.

Glanville, Ranulph. 2002. "Second Order Cybernetics." *Systems Science and Cybernetics* 3:59–85.

Glassie, Henry. 1999. *Material Culture.* Bloomington: Indiana University Press.

Goffman, Erving. 1956. *The Presentation of Self in Everyday Life.* New York: Anchor Books.

Goldman, Robert, and Stephen Papson. 1998. *Nike Culture: The Sign of the Swoosh*, Core Cultural Icons. London: SAGE.

Greenhouse, Steven. 1997. "Nike Shoe Plant in Vietnam Is Called Unsafe for Workers." *New York Times*, November 8, 1997. https://www.nytimes.com/1997/11/08/business/nike-shoe-plant-in-vietnam-is-called-unsafe-for-workers.html.

Greer, Betsy. 2008. *Knitting for Good: A Guide to Creating Personal, Social, and Political Change, Stitch by Stitch.* Boston: Trumpeter Books.

Greer, Betsy. 2017. "Why the World Needs This Project." You Are So Very Beautiful, May 20, 2017. http://youaresoverybeautiful.com/2017/05/20/why-the-world-needs-this-project/.

Groth, Camilla. 2017. "Making Sense through Hands. Design and Craft Practice Analyzed as Embodied Cognition." PhD diss., School of Arts, Design and Architecture, Aalto University.

Gygax, Gary. 1974. *Dungeons and Dragons.* Lake Geneva: TSR.

Halper, Vicki, and Diane Douglas, eds. 2009. *Choosing Craft: The Artist's Viewpoint.* Chapel Hill: University of North Carolina Press.

Hansen, James R. 1995. *SP-4308 Spaceflight Revolution.* Washington, DC: NASA Langley Research.

Haraway, Donna. 1988. "Situated Knowledges: The Science Question in Feminism and the Privilege of Partial Perspective." *Feminist Studies* 14 (3): 575–599.

Haraway, Donna J. 2016. *Staying with the Trouble: Making Kin in the Chthulucene.* Durham, NC: Duke University Press.

Hartman, Saidiya. 2008. "Venus in Two Acts." *Small Axe* 12 (2): 1–14.

Hayles, N. Katherine. 2017. *Unthought: The Power of the Cognitive Nonconscious*. Chicago: University of Chicago Press.

Holland, Dorothy, William Lachicotte Jr., Debra Skinner, and Carole Cain. 1998. *Identity and Agency in Cultural Worlds*. Cambridge, MA: Harvard University Press.

Honnet, Cedric, Hannah Perner-Wilson, Marc Teyssier, Bruno Fruchard, Juergen Steimle, Ana C. Baptista, and Paul Strohmeier. 2020. "PolySense: Augmenting Textiles with Electrical Functionality Using In-Situ Polymerization." In *Proceedings of the 2020 CHI Conference on Human Factors in Computing Systems*, 1–13. New York: ACM. https://doi.org/10.1145/3313831.3376841.

Hui, Yuk. 2016. *On the Existence of Digital Objects*. Minneapolis: University of Minnesota Press.

Ingold, Tim. 2009. "The Texility of Making." *Cambridge Journal of Economics* 34:91–102.

Ingold, Tim. 2013. *Making: Anthropology, Archaeology, Art and Architecture*. New York: Routledge.

Ingold, Tim, and Elizabeth Hallam. 2007. "Creativity and Cultural Improvisation: An Introduction." In *Creativity and Cultural Improvisation*, edited by Elizabeth Hallam and Tim Ingold, 1–24. Oxford: Berg.

Jacucci, Giulio. 2004. "Interaction as Performance." PhD diss., Faculty of Science, University of Oulu.

Jurkowski, Henryk. 1990. "The Mode of Existence of Characters of the Puppet Stage." In *The Language of the Puppet*, edited by Laurence R. Kominz and Mark Levenson, 21–37. Seattle: Pacific Puppetry.

Jurkowski, Henryk. 2014. "Terms in Puppetry and Their Linguistic and Research Context." In *Europske Odrednice Pojma Lutke i Strucno Lutkarsko Nazivje*, edited by Livija Kroflin, 7–19. Zagreb: Academy of Arts Osijek.

Kafai, Yasmin, and Kylie Peppler. 2014. "Transparency Reconsidered: Creative, Critical, and Connected Making with E-textiles." In *DIY Citizenship: Critical Making and Social Media*, edited by Matt Ratto and Megan Boler, 179–188. Cambridge, MA: MIT Press.

Kaptelinin, Victor, and Bonnie A. Nardi. 2006. *Acting with Technology: Activity Theory and Interaction Design*. Cambridge, MA: MIT Press.

Keller, Charles, and Janet Dixon Keller. 1994. "Thinking and Acting with Iron." In *Understanding Practice: Perspectives on Activity and Context*, edited by Seth Chaiklin and Jean Lave, 125–144. Cambridge: Cambridge University Press.

References

Kember, Sarah, and Joanna Zylinska. 2012. *Life after New Media: Mediation as a Vital Process*. Cambridge, MA: MIT Press.

Kester, Grant H. 2004. *Conversation Pieces: Community + Communication in Modern Art*. Berkeley: University of California Press.

Kinzer, Kacie J. 2010. Doll. US Design Patent. US D609,287 S.

Kleist, Heinrich von. (1811) 1982. "On the Puppet Theater." In *From an Abyss Deep Enough: Letters of Heinrich von Kleist with a Selection of Essays and Anecdotes*, edited by Philip B. Miller, 211–216. New York: E. P. Dutton.

Kosofsky, Leon J., and G. Calvin Broome. 1965. "Lunar Orbiter: A Photographic Satellite." Spring Convention of the Society of Motion Picture and Television Engineers, Los Angeles, March 28–April 2, 1965.

Latour, Bruno. 2005. *Reassembling the Social*. Oxford: Oxford University Press.

Laurel, Brenda. 1991. *Computers as Theatre*. Reading, MA: Addison-Wesley.

Lefebvre, Henri. 2004. *Rhythmanalysis: Space, Time and Everyday Life*. London: Continuum.

Lehmann, Hans-Thies. 2006. *Postdramatic Theatre*. Translated by Karen Juers-Munby. London: Routledge.

Leontiev, Aleksei Nikolaevich. 1978. *Activity, Consciousness, and Personality*. Englewood Cliffs, NJ: Prentice-Hall.

Levine, Faythe, and Cortney Heimerl. 2008. *Handmade Nation: The Rise of DIY, Art, Craft, and Design*. New York: Princeton Architectural.

Lichty, Patrick. 2009. "The Translation of Art in Virtual Worlds." *Leonardo Electronic Almanac* 16 (4–5). https://leonardo.info/LEA/DispersiveAnatomies/DA_lichty.pdf.

Lucie-Smith, Edward. 1981. *The Story of Craft: The Craftman's Role in Society*. Oxford: Phaidon.

Malafouris, Lambros. 2008. "At the Potter's Wheel: An Argument for Material Agency." In *Material Agency: Towards a Non-Anthropocentric Approach*, edited by Carl Knappett and Lambros Malafouris, 19–37. New York: Springer.

Manovich, Lev. 2001. *The Language of New Media*. Cambridge, MA: MIT Press.

Mattes, Eva, & Franco Mattes. 2007. "Nothing Is Real, Everything Is Possible." 0100101110101101.org. Accessed September 18, 2021. https://0100101110101101.org/press/2007-07_Nothing_is_real.html.

Maturana, Humberto R., and Francisco J. Varela. 1992. *The Tree of Knowledge: The Biological Roots of Human Understanding*. Rev. ed. Boston: Shambhala.

McKenzie, Jon. 2001. *Perform or Else: From Discipline to Performance*. New York: Routledge.

McLuhan, Marshall. (1987) 1997. *Understanding Media: The Extension of Man*. New York: McGraw-Hill.

Mellis, David A., Sam Jacoby, Leah Buechley, Hannah Perner-Wilson, and Jie Qi. 2013. "Microcontrollers as Material: Crafting Circuits with Paper, Conductive Ink, Electronic Components, and an 'Untoolkit.'" In *Proceedings of the 7th International Conference on Tangible, Embedded and Embodied Interaction*, 83–90. New York: ACM.

microRevolt. n.d. "Nike Blanket Petition." microRevolt. Accessed October 1, 2021. https://www.microrevolt.org/petition_overflow.html.

MoMA. 2011. "Tweenbot." MoMA. Accessed January 22, 2021. https://www.moma.org/interactives/exhibitions/2011/talktome/objects/146369/.

Mumford, Lewis. 1934. *Technics and Civilization*. New York: Harcourt, Brace.

Murray, Janet H. 1997. *Hamlet on the Holodeck: The Future of Narrative in Cyberspace*. Cambridge, MA: MIT Press.

MCA/Universal Pictures. 1966. *Around the Moon: Orbiter Camera Ready to Film Landing Sites*. New York: MCA/Universal Pictures.

Ninsve. 2007. "A New Performance by Franco and Eva Mattes!" *Nina & Plurabelle on 1st and 2nd Life Art*, March 20, 2007. http://pluraonart.blogspot.com/2007/03/new-performance-by-eva-and-franco.html.

Norman, Donald A. 2005. "Human-Centered Design Considered Harmful." *interactions* 12 (4): 14–19.

Oxman, Neri. 2007. "Digital Craft: Fabrication Based Design in the Age of Digital Production." In *Workshop Proceedings for Ubicomp 2007: International Conference on Ubiquitous Computing*, 534-538. Innsbruck: University of Innsbruck.

Papert, Seymour, and Idit Harel. 1991. *Constructionism*. Norwood, NJ: Ablex.

Paterson, Elaine Cheasley, and Susan Surette, eds. 2015. *Sloppy Craft: Postdisciplinarity and the Crafts*. London: Bloomsbury.

Peters, John Durham. 2015. *The Marvelous Clouds: Toward a Philosophy of Elemental Media*. Chicago: University of Chicago Press.

Phelan, Peggy. 1993. *Unmarked: The Politics of Performance*. London: Routledge.

Piris, Paul. 2014. "The Co-presence and Ontological Ambiguity of the Puppet." In *The Routledge Companion to Puppetry and Material Performance*, edited by Dassia N. Posner, Claudia Orenstein, and John Bell, 30–43. Florence: Routledge.

PopTechMedia. 2010. "Kacie Kinzer: Robot Love." YouTube video, 7:26, posted by PopTechMedia, February 17, 2010. https://www.youtube.com/watch?v=1mf8fXL9zGw.

Posner, Dassia N., Claudia Orenstein, and John Bell, eds. 2014. *Routledge Companion to Puppetry and Material Performance*. Florence: Routledge.

Press, Mike. 2007. "Handmade Futures: The Emerging Role of Craft Knowledge in Our Digital Culture." In *NeoCraft: Modernity and the Crafts*, edited by Sandra Alfoldy, 249–267. Halifax: Press of the Nova Scotia College of Art and Design.

Puig de la Bellacasa, María. 2017. *Matters of Care: Speculative Ethics in More Than Human Worlds*. Minneapolis: University of Minnesota Press.

Pye, David. 1968. *The Nature and Art of Workmanship*. Bethel, CT: Cambium.

Quake III Arena. 1999. *Win PC*. Mesquite, TX: Id Software.

Ratto, Matt. 2011. "Critical Making: Conceptual and Material Studies in Technology and Social Life." *Information Society: An International Journal* 27 (4): 252–260.

Ratto, Matt, and Garnet Hertz. 2019. "Critical Making and Interdisciplinary Learning: Making as a Bridge between Art, Science, Engineering and Social Interventions." In *The Critical Makers Reader: (Un)Learning Technology*, edited by Loes Bogers and Letizia Chiappini, 16–28. Amsterdam: Institute of Network Cultures.

Richards, Mary Caroline. 1966. *Centering*. Middletown, CT: Weslyean University Press.

Rinzler, Ralph, and Robert Sayers. 1980. *The Meaders Family: North Georgia Potters*. Washington, DC: Smithsonian Institution.

Risatti, Howard. 2007. *A Theory of Craft: Function and Aesthetic Expression*. Chapel Hill: University of North Carolina.

Roberts, David. 1990. *Mad Dog McCree*. Albuquerque: Atari/American Laser Games.

Rosner, Daniela. 2018. *Critical Fabulations: Reworking the Methods and Margins of Design*. Cambridge, MA: MIT Press.

Rosner, Daniela K., Samantha Shorey, Brock R. Craft, and Helen Remick. 2018. "Making Core Memory: Design Inquiry into Gendered Legacies of Engineering and Craftwork." In *Proceedings of the 2018 CHI Conference on Human Factors in Computing Systems*, 1–13. New York: ACM. https://doi.org/10.1145/3173574.3174105.

Ruskin, John. 1853. *The Stones of Venice. Vol 2. The Sea-stories*. London: Smith, Elder.

Salter, Chris. 2010. *Entangled: Technology and the Transformation of Performance*. Cambridge, MA: MIT Press.

Sassaman, Kenneth E. 1993. *Early Pottery in the Southeast: Tradition and Innovation in Cooking Technology*. Tuscaloosa: University of Alabama Press.

Sawyer, R. Keith. 2012. *Explaining Creativity: The Science of Human Innovation.* Oxford: Oxford University Press.

Schechner, Richard. 2002. *Performance Studies: An Introduction.* 2nd ed. New York: Routledge.

Schechner, Richard. 2003. *Performance Theory.* New York: Routledge.

Schön, Donald A. 1987. *Educating the Reflective Practitioner: Toward a Design for Teaching and Learning in the Professions.* San Francisco: Jossey-Bass.

Sennett, Richard. 2008. *The Craftsman.* New Haven, CT: Yale University Press.

Shindler, Kelly. 2010. "Life after Death: An Interview with Eva and Franco Mattes." *Art21 Magazine*, May 28, 2010. http://magazine.art21.org/2010/05/28/life-after-death-an-interview-with-eva-and-franco-mattes/.

Simondon, Gilbert. 1992. "The Genesis of the Individual." In *Incorporations*, edited by Jonathan Crary and Sanford Kwinter, 297–319. New York: Zone Books.

Simondon, Gilbert. (1958) 2017. *On the Mode of Existence of Technical Objects.* Translated by Cecile Malaspina and John Rogove. Minneapolis: Univocal.

Turkle, Sherry. 1984. *The Second Self: Computers and the Human Spirit.* New York: Simon & Schuster.

Turner, Victor. 1985. *On the Edge of a Bush: Anthropology as Experience.* Tucson: University of Arizona Press.

Vannini, Phillip, ed. 2015. *Non-representational Methodologies: Re-envisioning Research.* New York: Routledge.

Virilio, Paul. 1995. "Speed and Information: Cyberspace Alarm!" *CTheory.* http://www.ctheory.com/a30-cyberspace_alarm.html.

Vygotsky, Lev. 1930. *Mind and Society.* Cambridge, MA: Harvard University Press.

Welsch, Wolfgang. 2012a. *Homo mundanus: Jenseits der anthropischen Denkform der Moderne.* Weilerswist, Germany: Velbrück.

Welsch, Wolfgang. 2012b. *Mensch und Welt: Philosophie in evolutionärer Perspektive.* Munich: Beck.

Westecott, Emma. 2013. "Independent Game Development as Craft." *Loading* 7 (11): 78–91.

Wiberg, Mikael. 2018. *The Materiality of Interaction: Notes on the Materials of Interaction Design.* Cambridge, MA: MIT Press.

Wiener, Norbert. (1961) 2019. *Cybernetics: Control and Communication in the Animal and the Machine.* Cambridge, MA: MIT Press.

References

Wilson, Frank R. 1999. *The Hand: How Its Use Shapes the Brain, Language, and Human Culture*. New York: Vintage Books.

Ziemann, Benjamin. 2013. *Contested Commemorations: Republican War Veterans and Weimar Political Culture*, Studies in the Social and Cultural History of Modern Warfare. Cambridge: Cambridge University Press.

Zoran, Amit. 2016. "A Manifest for Digital Imperfection." *XRDS: Crossroads; the ACM Magazine for Students* 22 (3): 22–27.

Zoran, Amit, and Leah Buechley. 2010. "Hybrid Reassemblage: An Exploration of Craft, Digital Fabrication and Artifact Uniqueness." *Leonardo* 46 (1): 4–10.

Index

3D print, 13, 117, 158, 161
7000 Oaks (project), 15, 54–59, 61, 63, 73, 79, 106, 200, 202. *See also* Beuys, Joseph
7000 Oaks (reenactment) (project), 58–61, 63, 73, 105. *See also* Mattes, Eva & Franco
Aarseth, Espen, 64, 86, 87
Abramović, Marina, 58, 77–78, 82
Action network, 30, 35, 40–41, 86, 156
Actor network theory, 36
Actual, 82–86, 106, 133. *See also* Kaprow, Allan; Schechner, Richard
Adamson, Glenn, 16, 110, 115, 120, 122, 125–126, 133–134, 150, 160
Adversarial, 54, 97–99, 107, 113. *See also* DiSalvo, Carl
Agential cut, 144–145, 154. *See also* Barad, Karen
AI (artificial intelligence), 8, 79, 86–87, 89–90, 197
AIDS quilt (project), 120
Albers, Annie, 143
Ampex FR-900, 23, 25. *See also* Lunar Orbiter 1; Lunar Orbiter Image Recovery Project
Android (operating system), 102–105
Anonym, 69–70, 73, 103–105
Ansari, Ava, 100–105, 198. See also *Subway* (project)
Anthropocene, 2, 13, 22–23, 25

Anthropology, 12–13, 62, 82, 129, 175, 189
Apollo (space program), 19, 21, 195
App (mobile), 102–105, 151, 173
Augmented Reality (AR), 12, 25, 99, 102, 198
Auslander, Philip, 15, 64, 75–76, 105
Autopoietic feedback loop, 77–79, 90, 107, 143. *See also* Fischer-Lichte, Erika

Barad, Karen. *See also* Material agency
  agency, 27, 37, 43, 105
  intra-action, 27, 28, 38–39, 48, 144–145
  new materialism, 5, 10, 14
  performative, 76, 78
  phenomena, 149
Bell, John, 15, 91, 94, 96
Bennett, Jane, 5, 10, 14, 39, 43, 48
Bergson, Henri, 36–37, 41
Beuys, Joseph, 15, 54–61, 63, 73, 79, 83, 107, 117, 160, 168, 200. See also 7000 Oaks (project)
Bigmeat family, 180
Blendie (project), 70–73, 98–99, 198. *See also* Dobson, Kelly
Boal, Augusto, 64
Bolter, Jay, 43
Borgmann, Albert, 7, 24–25, 69, 73, 130, 182

Bourriaud, Nicolas, 36, 120
Branding, 128, 130, 169, 172–173
Brick Bros. Circus, The, 94, 96, 97
Buechley, Leah, 119, 155, 158–159
Burrison, John, 16, 132, 168, 176–186

Care, 67, 120, 142, 163, 165, 201
Ceramic, 16, 111, 114–117, 127, 131, 144–145, 157–159, 179–183
Changes (project), 145–146
Codependent, 38, 81–82, 93, 122, 135
Coemergence, 28, 31, 35, 37, 40, 110, 126, 136
Coexistence, 5, 40, 73, 117, 122, 125, 163
Cold War, 21, 63
Commodity, 43, 69–70, 72, 115, 130–131, 183. *See also* Borgmann, Albert
Community-based practice, 16, 168, 184–190, 193, 201
Concretization, 45, 49–50, 63, 149, 198
Craft
  definition, 118–121
  digital impact, 117–120
  revival, 15, 118, 120, 176
  social construction, 118–119, 123
  supplemental, 115, 126
Craftivism, 16, 120, 168, 191–192, 195, 201. *See also* Greer, Betsy
Craven, Mike, 182, 184
Critical Design, 153–154
Critical fabulation, 196–197, 201
Critical making, 16, 110, 152–156, 162, 165. *See also* Hertz, Garnet; Ratto, Matt
Cubitt, Sean, 3, 27
Cybernetics, 32–33, 46, 50, 69, 128
Cyborg, 6

Damage, 147, 159–160
Dancing by Myself in Public (project), 100–103
Dave the Potter, 181
Davidson, Carolyn, 169, 174

Detachment, 24–25, 40, 45–46, 73, 93, 133
Device paradigm, 69. *See also* Borgmann, Albert
Dewey, John, 141, 153
Differentiation (human/ nonhuman)
  dialogical, 35–36
  vs milieu, 47–50, 196
  self, 73
DiSalvo, Carl, 97–98
Disembodiment, 22, 25, 75, 161–162
Diversity, 2, 110, 123–125, 128–131
Dix, Otto, 111–116, 118
Dobson, Kelly, 70–73, 178. See also *Blendie* (project)
Dormer, Peter, 110, 150
Dorson, Richard, 16, 168, 175–177, 184–185
Dungeons and Dragons (game system), 145, 147–148

Earthrise, 14, 18–19, 21–23, 25–26, 29
Education
  approach, 13–14, 16
  collaboration, 163
  craft, 182–184, 187
  and Critical making, 110, 152–156, 162, 165
  hands on, 141, 143
Embodied
  cognition, 31, 123
  craft, 125–126, 132, 174, 196
  expression, 63, 157–159, 182
  interaction design, 137, 161
  puppets, 94
  robot, 98–99
  view, 22
Ergodic, 86–87, 90
Escobar, Arturo, 188, 191
e-textile, 137, 154
Evolution
  approach, 10, 12, 41, 49, 51, 67, 98, 107, 118

Index

cognitive, 18, 32–38, 40–41, 44, 87, 122–123, 150
digital, 6, 76
folk, 79, 175, 185–186, 198–201
material, 132, 135
Experience design, 4, 82
Eyebeam, 99–100

Fabric
 material, 92, 99, 138, 163–164, 172, 177
 social, 62, 83, 130, 177, 187, 198
Fabrication, 8, 11, 16, 110, 118–119, 127, 131, 133–134, 157–160, 187
Failure, 142–144, 161
Faught, Josh, 163–164
Feminist, 5, 120, 196
Film, 19–20, 26, 63–64, 74, 100
Fischer-Lichte, Erika, 15, 77–78, 90, 143
Fluids (project), 84–85. See also Kaprow, Allan
Folk. *See also* Folk potter
 art, 99, 176, 178–179, 182, 184
 education, 177, 184, 187
 folklife, 16, 168, 175–178, 185, 201
 media, 176–178
Folk potter, 16, 131, 168, 180–185, 194
Folk Pottery Museum of Northern Georgia, 183

Glassie, Henry, 129–130, 183
Goffman, Erving, 65–67, 74
Greer, Betsy, 16, 120, 168, 190–195, 201. *See also* Craftivism; *You Are So Very Beautiful* (project)

Handmade, 121, 127–128, 137, 157, 164, 177–178, 183–184, 192, 195
Haraway, Donna, 10, 12, 22, 30, 35, 43, 48–49, 91, 144, 156, 200
Hayles, Katherine N., 29, 31–32, 35
Hertz, Garnet, 152–156
hitchBOT (project), 80–82, 84, 90

Homo mundanus, 32, 37, 40. *See also* Welsch, Wolfgang
Humancentric, 2–3, 5, 14, 75, 77, 86, 88, 168, 198–199, 201
Hybrid
 craft, 110, 119, 155, 201
 reassemblage, 16, 131, 158–162, 165, 201 (*see also* Zoran, Amit)

Identity, 2, 8, 29, 67, 90, 98, 124–128, 130–131, 143, 149, 157, 169, 186–187
Imperfection, 159–160, 194
Improvisation, 16, 110, 128, 140, 142–143, 147, 149, 151, 155, 201
Individuation
 craft, 121, 139, 151
 evolution, 34, 37, 40, 46–50, 110, 137 (*see also* Simondon, Gilbert)
 expression, 40, 50, 175
 with material, 134–137, 139, 144, 149, 198
 need, 107, 110, 122, 124–125, 174
 self, 4, 143, 198
Industrial Revolution, 24, 63, 118, 127, 160
Ingold, Tim, 10, 16, 28, 141–142
Interaction design, 15, 44, 64–65, 78, 86, 137, 151, 185–186
Internet, 7, 25, 43, 58, 61, 63–64, 85, 119, 197
Intra-action, 18, 26–28, 35, 39, 41, 48, 50, 107, 145, 175, 200. *See also* Barad, Karen
ioq3aPaint (project), 88–90. *See also* Oliver, Julian

Jewelry, 122–123, 125, 149

Kaprow, Allan, 82, 84–85. See also *Fluids* (project); Actual
Keller, Charles & Janet, 141–142
Kember, Sarah, 5, 9, 12–14, 27–30, 37

Kinzer, Kacie, 79–80. See also *Tweenbot*
Kleiman, Molly, 99–100
knitPro (project), 173, 189
Knitting, 170, 172–173, 186, 190
Kodak, 19–20, 30

Landrum, Abner, 180–181
Latour, Bruno, 10, 38
Lehmann, Hans-Thies, 106
Leontiev, Aleksei, 79
Lived material culture, 16, 168, 183–185, 188–190, 193, 201
Liveness, 54, 64, 76. See also Auslander, Philip
Lucie-Smith, Edward, 110, 119, 123
Lunar Orbiter 1, 14, 18–25, 29, 34, 41, 48, 106
Lunar Orbiter Image Recovery Project, 23–24

Machine, 63, 69
 computer, 8, 19, 23
 digital, 44, 62, 76, 161, 196
 encountering, 70–72, 96
 industrial, 118–119, 127–128
 learning, 197
 as mediator, 44–45
 and performance, 74, 88, 90
 robotic, 69, 79, 98–99 (*see also* Dobson, Kelly)
 tape, 23, 25
 textual, 64, 86
Mad Dog McCree (video game), 82, 84–85
Maker culture, 161, 187
Making Core Memory (project), 195–197. See also Rosner, Daniela
Marionette, 91–94. See also Puppet
Mask, 96, 123
Material agency
 Barad, Karen, 26–27, 39, 105
 digital, 42–43
 and human agency, 39, 200

performance, 54, 70–71, 86, 107
personal fabrication, 16, 198
Material culture
 computer, 196
 craftivism, 168
 disconnect, 24, 131
 diversity, 110
 field, 11, 129
 folk, 16, 124, 168, 176, 178, 183–185, 188–190, 201
Material performance, 15, 54, 91, 94, 99
Mattes, Eva & Franco, 54, 58–62, 74
Maturana, Humberto, 34–35, 38, 188
Mazza, Cat, 16, 168, 170–174, 178, 189–190, 192, 198, 201. See also *Nike Blanket Petition* (project)
McKenzie, Jon, 63, 105
McLuhan, Marshall, 26
Meaders, Lanier, 180, 183
Media, 30
 designer, 4, 6, 9–10, 36, 50, 198, 202
 ecology, 4–5, 7, 16, 25, 168, 191
microRevolt, 170–173. See also Mazza, Cat
Milieu, 15, 18, 47–48, 50–51, 74, 117, 131, 155, 186, 198. See also Simondon, Gilbert
Morris, William, 42–44, 48, 118, 120, 160
Murray, Janet, 5, 44, 64, 82

NASA, 21, 23
Need
 biological, 32, 201
 care, 163
 craft, 16, 110, 117, 121–126, 133, 139, 143, 171, 194
 human, 3, 4, 7, 10, 14, 30, 48, 67, 79, 90, 107, 176, 180–187, 200
 material, 4, 35, 107, 110, 122, 155
 technical, 19, 87, 121, 124, 143
New materialism, 5, 11, 14, 18, 26–27, 33, 38, 76

Index

Nike (company), 169–172, 174
Nike Blanket Petition (project), 16, 168–174, 178, 189, 192, 198, 201–202. *See also* Mazza, Cat
Nonconscious cognition, 31. *See also* Hayles, Katherine N.
Norman, Don, 79

Oliver, Julian, 88–89. *See also ioq3aPaint* (project)
Orenstein, Claudia, 15, 91, 94

Performance (art), 65. *See also 7000 Oaks* (project); *Rhythm 0* (project)
Performance studies, 15, 54, 62–64, 99, 105
Perner-Wilson, Hannah, 137–140, 155, 162, 165
Peters, John Durham, 12, 26, 29, 83
Phenomena, 26–27, 149, 175
Photosynthesis, 30, 79
Player
  game, 64, 82, 86–87, 89, 105, 145–148, 161
  nonhuman, 89, 106, 148
PlayStation, 6
Political
  cold war, 19, 22
  craft, 120–121, 123, 163, 175, 181, 191
  design, 98, 188. *See also* DiSalvo, Carl
  Germany, 55, 114, 175. *See also* Beuys, Joseph
  performance, 67, 98–100, 104
  puppetry, 98
Posner, Dassia, 15, 91, 94
Postdramatic, 106
Posthuman, 27, 78, 156, 201
Primal Clay (project), 145–148, 202
Procedural, 5, 7, 42, 44–45, 88, 90, 201
Puig de la Bellacasa, María, 163
Puppet, 5, 15, 54, 91–99, 106–107, 137, 151, 200. *See also* Marionette
Pye, David, 125, 128–129, 142–144, 160

Quake III (video game), 88
Quitmeyer, Andrew, 101–105, 198. *See also Subway* (project)

Ratto, Matt, 16, 152–156. *See also* Critical making
Rhythm 0 (project), 78–80, 82, 84. *See also* Abramović, Marina
Risatti, Howard, 16, 110, 121–125, 171
Risk, 3, 82, 97, 142–144, 158–161, 164
Ritual, 62, 82–83
Robot, 15, 65–66, 74, 79–82, 85–88, 98, 107, 154, 197, 200. *See also hitchBot* (project); *Tweenbot*
Rosner, Daniela, 5, 195–196

Schechner, Richard, 62–63, 82–84, 88, 99–100, 106
Schön, Donald, 142
Second Life, 54, 58–62, 73
Sennett, Richard, 110, 123, 134–135, 149–150
Simondon, Gilbert, 14–15, 18, 42–49, 63, 74, 110, 117, 123, 135–137
Skill, 69, 119, 125–126, 144, 152, 162–163
Sloppy craft, 163, 165. *See also* Faught, Josh
Smith, David Harris, 80. *See also hitchBot* (project)
Social media, 5, 8, 25, 41, 61, 192, 194, 198
Social sculpture, 15, 54–57, 61–62, 73, 85, 105, 107, 116, 130, 168, 198, 202. *See also* Beuys, Joseph
Spaceship Earth, 21, 40
Speculation, 153, 196
Stradivari, 127, 157
Strawberry Thief, 42–44. *See also* Morris, William
String figures, 49, 91–92. *See also* Haraway, Donna

Subway (project), 15, 54, 99–106, 151, 160, 198, 202. *See also* Ansari, Ava; Quitmeyer, Andrew

Tacit (knowledge), 143, 161
Tactical media, 9–10, 174, 189
Television, 38, 46, 64, 74–75
Theater of the Oppressed, 64. *See also* Boal, Augusto
Tool
  digital, 65, 119, 133, 161, 168, 173
  hybrid, 137–139, 155, 160, 187
  maker, 13, 46, 79, 117, 123–124, 129, 135, 143, 150, 177
  media, 5, 24, 26, 195–196
Tool, Ehren, 15, 110–119, 125–130, 144, 160, 162, 194, 198, 200–201
Transformation, 63, 77, 82–83, 85–86, 106, 132, 149
Turpin, Steve, 183–184, 187
Tweenbot, 79–83, 90, 198
Twitter, 81, 192, 194

Ubiquitous computing, 2, 6, 15, 25, 64

Varela, Francisco, 34–35, 38, 188
Variation within tradition, 16, 168, 183–185, 188–190, 193, 201
Video game
  activity, 15, 45, 84, 86
  craft, 161, 201
  discursive encounter, 145
  distance, 8
  media, 26, 198
  performance, 68, 74, 85, 87–89
  textual machine, 64–65, 86
  transformation, 82
Virtual world, 6, 58, 60–62
Vitalism, 36
Vital media definition, 41, 49–51
von Kleist, Heinrich, 91, 93–94
Vygotsky, Lev, 79

Welsch, Wolfgang, 14, 18, 32–37, 40–41, 66, 88, 123
Wheel (pottery), 30, 131, 144, 180, 183. *See also* Ceramic
Wiener, Norbert, 46, 69
Workshop
  craft, 127, 131, 135, 143, 183, 192–193, 202
  educational, 153–155, 171, 189, 195. *See also* Critical making
  Subway (project), 99–100, 103
World War
  WWI, 113, 114, 118
  WWII, 19, 55, 113–114, 143, 157

You Are So Very Beautiful (project), 192–193, 194, 201. *See also* Greer, Betsy

Zeller, Frauke, 80. See also *hitchBot* (project)
Zoran, Amit, 16, 110, 119, 131, 156–165, 201
Zylinska, Joanna, 5, 9, 12–14, 27–30, 37